CRITICAL ISSUES IN URBAN ECONOMIC DEVELOPMENT

Publications in the Inner Cities Research Programme Series

1 Overview volume
Urban Economic Adjustment and the Future of British Cities: Directions for Urban Policy
Victor A. Hausner, Policy Studies Institute, London

2 Summary volume on five city studies
Urban Economic Change: Five City Studies
Members of the five research teams and the National Director

3 Bristol region study
Sunbelt City? A Study of Economic Change in Britain's M4 Growth Corridor
Martin Boddy, John Lovering, and Keith Bassett

4 Clydeside conurbation study
The City in Transition: Policies and Agencies for the Economic Regeneration of Clydeside
edited by W. F. Lever and Chris Moore

5 London metropolitan region study
The London Employment Problem
Nick Buck, Ian Gordon, and Ken Young, with John Ermisch and Liz Mills

6 West Midlands study
Crisis in the Industrial Heartland: A Study of the West Midlands
Ken Spencer, Andy Taylor, Barbara Smith, John Mawson, Norman Flynn, and
Richard Batley

7 Newcastle metropolitan region study
Economic Development Policies: An Evaluative Study of the Newcastle Metropolitan Region
Fred Robinson, John Goddard and Colin Wren

8 Special research studies
Critical Issues in Urban Economic Development, vol. I
edited by Victor A. Hausner, Policy Studies Institute, London

9 Special research studies
Critical Issues in Urban Economic Development, vol. II
edited by Victor A. Hausner, Policy Studies Institute, London

10 Special research study
The New Economic Roles of UK Cities
Iain Begg and Barry Moore, Department of Applied Economics, University of
Cambridge

11 Collective volume
Economic Change in British Cities
Members of the five research teams and the National Director

CRITICAL ISSUES IN URBAN ECONOMIC DEVELOPMENT

Volume II

Edited by
VICTOR A. HAUSNER

CLARENDON PRESS · OXFORD

1987

Oxford University Press, Walton Street, Oxford OX2 6DP

Oxford New York Toronto
Delhi Bombay Calcutta Madras Karachi
Petaling Jaya Singapore Hong Kong Tokyo
Nairobi Dar es Salaam Cape Town
Melbourne Auckland

and associated companies in
Beirut Berlin Ibadan Nicosia

Oxford is a trade mark of Oxford University Press

Published in the United States
by Oxford University Press, New York

British Library Cataloguing in Publication Data

Critical issues in urban economic development.—(Inner cities research programme)   Vol. 2
1. Cities and towns—Great Britain
2. Urban economics
I. Hausner, Victor   II. Series
330.941'0858   HT321
ISBN 0-19-823268-3

Library of Congress Cataloging in Publication Data (Revised for vol. 2)

Critical issues in urban economic development.
(Inner cities research programe; 8   )
Includes indexes.
1. Community development, Urban—Great Britain.
2. Urban policy—Great Britian.   3. Great Britain—
Economic conditions—1945–   I. Hausner, Victor A.
II. Series.
HN400. C6C75   1986   307.1'4'0941   86–11342
ISBN 0-19-823266-7 (v. 1)
ISBN 0-19-823268-3 (v. 2)

Set by Promenade Graphics Ltd, Cheltenham
Printed and bound in Great Britain by
Biddles Ltd, Guildford and King's Lynn

FOREWORD

In 1982 the Environment and Planning Committee of the Economic and Social Research Council (SSRC at the time) initiated a three year comparative research programme to examine inner city problems within the broad context of major structural and spatial changes occurring in Great Britain. The programme was developed by the then SSRC Inner Cities Working Party, chaired by Professor Peter Hall, and subsequently the Executive Panel on the Inner Cities, chaired by Professor Gordon Cameron. The proposal for the research was originally described in 'A Research Agenda' (chapter 8) of *The Inner City in Context* (ed. Peter Hall, Heinemann, 1981). The purpose of the programme was to examine the processes of urban change, the effects on urban socio-economic welfare, and the prospects, constraints, and requirements for more successful urban adjustment to structural change. The programme arose from concerns with the urban problems of economic decline, labour market imbalances, social distress, and the effectiveness of public policies in addressing these problems. It was hoped that the programme's findings would be useful for the improvement of public policies to strengthen urban economies (that is, foster growth, employment, and competitiveness), alleviate the distress caused by change and improve the conditions of distressed inner city areas and deprived urban residents.

In practice, the overall programme focused on the economic aspects of urban change. Specifically, the programme sought to identify the key factors affecting urban economic change and to describe and explain the processes of local economic change. Secondly, it aimed to describe the consequences of change for the urban economy and the employment of its residents. Finally, the programme attempted to assess the effects of public policies on the process of change.

The programme was based on the idea that there is diversity in the economic performance of different urban and inner city areas in the UK and in their adjustment to external forces of change; for example, changes in business competition, technology, and residential patterns. A comparative examination of the nature, processes, and effects of economic change on different urban centres should help to clarify and explain the differences in the experiences of economic change among and within urban areas and to identify those factors (including public policies) which impede or facilitate urban and inner city adjustment to change: that is, economic growth, increased business competitiveness, employment generation, employment for the disadvantaged in urban labour markets, and the effectiveness of urban economic policies. The ESRC programme utilized a general framework of topics to assist individual urban studies and the comparative examination of urban economic change. The topics were: the changing nature of urban economic problems; the nature and causes of imbalances in urban labour markets; the unintended effects of central and local government policies on urban economic change, and the effectiveness of national and local urban economic policies; and the capabilities of local authorities to design and implement more effective economic development policies.

The Inner Cities Research Programme addressed these concerns through two avenues of work. The core of the programme was independent studies of four major urban centres of Great Britain which were selected by the SSRC as examples of the diversity of urban economic adjustment experiences: Glasgow and the Clydeside conurbation as an example of 'persistent economic decline'; Birmingham and the West Midlands conurbation as one of 'faltering growth'; the Bristol region as one of 'successful adaptation'; and two areas in London, one (Greenwich, Southwark, and Lewisham) exhibiting 'severe problems', the other (Brent, Ealing, and Hounslow) exemplifying more 'successful adaptation'. These were largely secondary research studies using existing data and analysing existing research. The studies were conducted by independent research teams with distinctive approaches and concerns, but linked by the overall programme's objectives and general framework of topical concerns, dialogue, information exchange, and some common data and analysis. The different research teams decided on the particular approach, subjects, and hypotheses which they considered most relevant to an understanding of economic change in their study areas; and critically examined the characterizations of those areas.

To these four initial studies a fifth study of the Newcastle metropolitan region was added, funded by the Department of the Environment, in order to broaden the sample of urban areas and focus particularly on an assessment of the relationship, impacts, and effectiveness of central and local government urban and regional economic development policies on an economically distressed city region that was a long-term recipient of government assistance.

In addition to these five core studies a number of smaller 'cross-cutting' studies were conducted by various researchers in order to provide a national statistical framework for the five city studies and uniform comparative data on the five city study areas, to examine in greater detail important aspects of urban economic change in a broader sample of urban areas, to explore the effects on change of important public policies, to provide a comparative international perspective, and to increase the general relevance of a case-study-based research programme. A list of the publications and their authors, resulting from all of these elements of the programme, appears at the beginning of this book.

In order to enhance the relevance of the research programme to public policy issues and communication with central government policy-makers, discussions were held with officials of three government agencies: the Department of the Environment, the Department of Trade and Industry, and the Manpower Services Commission. These agencies also assisted the programme through the provision of data, special analyses, and the co-operation of their regional offices. The individual city and cross-cutting studies involved extensive local contacts with government officials, representatives of business, labour and voluntary organizations, and other researchers and analysts.

The five city studies were conducted by members of research teams at the following institutions: the Department of Social and Economic Research at the University of Glasgow; the Institute of Local Government Studies and the Centre for Urban and Regional Studies at the University of Birmingham; the School for Advanced Urban Studies and the Department of Geography at the University of Bristol; the Policy Studies Institute in London and the Urban and Regional Studies Unit at the University of Kent; and the Centre for Urban and Regional Development Studies at the University of Newcastle upon Tyne.

Professor Brian Robson, chairman of the ESRC Environment and Planning Committee, and Professor Noel Boaden and Paul McQuail, both members of the Committee, advised on the implementation of the programme. They were assisted by Dr Angela Williams, Senior Scientific Officer to the Committee. Members of the former SSRC Executive Panel discussed, reviewed papers, and advised on the research during the course of the programme. The programme also benefited from the advice and comments of other urban analysts.

London 1985

VICTOR A. HAUSNER
Director, ESRC Inner Cities Research Programme

CONTENTS

List of Contributors   x

List of Figures   xi

List of Tables   xii

1   Introduction
    Victor A. Hausner   1

2   Urban Economic Performance: A Comparative Analysis
    Harold Wolman   9

3   The Changing Economic Role of Britain's Cities
    Iain Begg and Barry Moore   44

4   The Beneficiaries of Employment Growth: An Analysis of the
    Experience of Disadvantaged Groups in Expanding Labour
    Markets
    Nick Buck and Ian Gordon   77

5   Local Employment and Training Initiatives in the National
    Manpower Policy Context
    Andrew A. McArthur and Alan McGregor   116

6   Housing Policies, Markets, and Urban Economic Change
    John Ermisch and Duncan Maclennan   160

Index   201

LIST OF CONTRIBUTORS

Iain Begg is a Research Officer in the Department of Applied Economics at the University of Cambridge.

Nick Buck is a Research Fellow in the Urban and Regional Studies Unit at the University of Kent at Canterbury.

John Ermisch is a Research Officer at the National Institute of Economic and Social Research, London. Formerly he was a Senior Research Fellow at the Policy Studies Institute, London, and he worked as a research economist at the US Department of Housing and Urban Development in Washington, DC.

Ian Gordon is Reader in Regional Studies and Director of the Urban and Regional Studies Unit at the University of Kent at Canterbury.

Victor A. Hausner at the time of writing was National Director of the ESRC Inner Cities Research Programme and Senior Visiting Fellow at the Policy Studies Institute in London. He is now Managing Partner of PA Cambridge Economic Consultants. From 1977 to 1981 he was Deputy Assistant Secretary for Economic Development Policy and Planning in the US Department of Commerce.

Andrew A. McArthur is a Research Fellow in the Department of Social and Economic Research at the University of Glasgow.

Alan McGregor is Senior Lecturer in Labour Economics in the Department of Social and Economic Research at the University of Glasgow. He has extensive experience researching aspects of urban employment and has worked as an adviser and consultant to a variety of public bodies on manpower and training policy issues.

Duncan Maclennan is the Director of the Centre for Housing Research at the University of Glasgow. He has published books and articles on urban, regional, and housing economics. Recently he has been an economic adviser on housing rehabilitation policy to Glasgow District Council, the Housing Corporation, and the Organization for Economic Co-operation and Development (Paris).

Barry Moore is Assistant Director of Research in the Department of Land Economy at the University of Cambridge, and a Fellow of Downing College.

Harold Wolman is Professor in the Department of Political Science at Wayne State University (Detroit, Michigan). At the time of writing he was Visiting Professor in the Department of Politics at the University of Salford (England).

LIST OF FIGURES

4.1 Path Diagram summarizing the Relationships involved in Disadvantage in the Labour Market   80
4.2 The Selected Areas of Employment Growth   84
6.1 City of Glasgow District: Private Sector Housing Areas   164
6.2 City of Glasgow District: House-price Deciles of Census Areas, 1983   166
6.3 Glasgow Housing and Redevelopment Areas   171
6.4 The GLC Area and the London Boroughs: The Percentage of Households in Council Housing, 1981   172
6.5 Some of the Repercussions of Housing Reinvestment   179

LIST OF TABLES

2.1 The Good and Poor Urban Performers considered in the Comparative Analysis   13–14

2.2 Employment Growth and Unemployment Performance in Britain, the USA, and Germany   15

2.3 Urban Economic Performance and Region in Britain, the USA, and Germany   16

2.4 Urban Economic Performance and Population Size in Britain, the USA, and Germany   18

2.5 Employment Growth and Unemployment Rate Performance by Size of Urban Area in Britain, the USA, and Germany   20–21

2.6 Economic Performance and Urban Function in the USA   22

2.7 Performance and Industrial Structure in Britain, the USA, Germany, and the Netherlands   23

2.8 Impact of Percentage of Total Employment engaged in Manufacturing in 1970 on Total Employment Change and Unemployment Rate Change in Britain, the USA, and Germany   24

2.9 Median Percentage Change in Employment by Sector in Britain, the USA, Germany, and the Netherlands   26

2.10 Economic Performance by Occupation in Britain and the USA   31

2.11 Urban Economic Performance of National Urban Systems in Britain, the USA, Germany, and the Netherlands   35

2.12 Extent of Variation in Urban Economic Performance by Country for Britain, the USA, and Germany   35

2.A1 Definitions of Performance Groups for Britain, the USA, and Germany   41

3.1 Proportions of Employment by Industry Group in Great Britain, 1971 and 1981   48

3.2 Employment by Sector based on Employment Exchanges, 1971   52

3.3 Employment by Sector based on Employment Exchanges, 1981   53

3.4 The Twenty Fastest-growing Cities in Great Britain, 1971–1981   55

3.5 Location Quotients for each of the Nine Industrial Groups for the Twenty Fastest-growing Cities as at 1971 and 1981   56

3.6 The Bottom Twenty Declining Cities in Great Britain, 1971–1981   57

3.7 Location Quotients for each of the Nine Industrial Groups for the Bottom Twenty Declining Cities as at 1971 and 1981   58

3.8   The Percentage Contribution of the Differential Growth and Structural Components to Actual Employment in each of the Twenty Fastest-growing and Twenty Slowest-growing Cities, 1971–1981   60

3.9   Results of Regression Analysis to Test for the Association of each of the Nine Industrial Sectors with City Employment Performance   63

3.A1 Employment in the Top and Bottom Twenty Cities in Industry Group A as a Percentage of Total Employment in 1971 and 1981   68

3.A2 Employment in the Top and Bottom Twenty Cities in Industry Group B as a Percentage of Total Employment in 1971 and 1981   69

3.A3 Employment in the Top and Bottom Twenty Cities in Industry Group C as a Percentage of Total Employment in 1971 and 1981   70

3.A4 Employment in the Top and Bottom Twenty Cities in Industry Group D as a Percentage of Total Employment in 1971 and 1981   71

3.A5 Employment in the Top and Bottom Twenty Cities in Industry Group E as a Percentage of Total Employment in 1971 and 1981   72

3.A6 Employment in the Top and Bottom Twenty Cities in Industry Group F as a Percentage of Total Employment in 1971 and 1981   73

3.A7 Employment in the Top and Bottom Twenty Cities in Industry Group G as a Percentage of Total Employment in 1971 and 1981   74

3.A8 Employment in the Top and Bottom Twenty Cities in Industry Group H as a Percentage of Total Employment in 1971 and 1981   75

3.A9 Employment in the Top and Bottom Twenty Cities in Industry Group J as a Percentage of Total Employment in 1971 and 1981   76

4.1   Regional Employment Change in Great Britain, 1971–1981   82

4.2   Percentage Employment Change in the Selected Growth Areas, 1971–1981   83

4.3   Percentage Change in Employment by Occupation in Growth Areas, 1971–1981   86

4.4   Unemployment Rates for Disadvantaged Groups, 1981   93

4.5   Percentage Unemployed in 1981 by Occupation in Previous Year   95

4.6   Percentage of Unemployed Seeking Work for more than Twelve Months, 1981   96

4.7 Logit Estimates for Probability of Males being Unemployed, 1981   100

4.8 Logit Estimates for Probability of Females being Unemployed, 1981   101

4.9 Occupational Shifts by Region and Conurbation, 1966–1971   106

4.10 Analysis of Variance in the Instability Characteristics of Jobs   108

4.11 Analysis of Variance for No Qualifications   109

5.1 Cash Expenditure (£m.) on Selected MSC Activities   122

5.2 Community Programme Filled Places and Size of Labour Force by Region   123

5.3 Percentage Distribution of YTS Entrants by Mode, 1985   126

6.1 Average Prices of Houses in Radial Bands round Central London: Three-bedroom Houses sold between October 1965 and March 1966   165

6.2 Percentage of Council Tenants in SEGs 1–5 in British Towns and Local Authority Areas, 1981   173

6.3 Percentage in Tenure, by SEG: Great Britain, 1982   181

6.4 Percentage in Tenure, by Income Group: Great Britain, 1982   181

6.5 Annual Movement in Individual Tenure Groups by Residents of the Manchester–Salford Inner area, 1978–1979 (Excluding Student Households)   182

6.6 The Impacts of Housing Rehabilitation Programmes in Glasgow   189

1

Introduction

Victor A. Hausner

This volume contains a series of essays on the findings of a second round of special 'cross-cut' research projects which were conducted as part of the Economic and Social Research Council (ESRC) Inner Cities Research Programme. This research was concerned with examining further several major themes of the overall programme. The first is the search for a better understanding of the effects of international and national structural economic change on urban economies and the identification of those factors which help to explain the diversity in economic performance among urban areas subject to similar external forces. The analysis of British urban economic change is extended here to an international comparison of urban economic performance in Britain and several other Western advanced industrial nations. Do the urban systems of advanced industrial nations share certain economic problems, and are the urban areas the locus of poor national economic performance? Is there diversity in urban economic experiences among industrial nations, what explains the diversity, and what distinguishes Britain's experience? The research places British urban economic experience in an international context.

It has been proposed for the USA that structural economic change, and particularly the rise of the service industries, is altering the economic functions of cities with significant effects on urban economic performance and resulting in a new hierarchy within the urban system (Noyelle and Stanback 1984). The relevance of this proposition to Britain is examined in an important study comparing the sectoral characteristics of the nation's largest urban areas. What is the functional composition of Britain's more economically successful cities, and are the growing economic functions locating away from the regional and urban centres of traditional industrial growth, thus changing the geographic distribution of economic activity in Britain and its urban economic hierarchy?

Another theme is labour-market processes and their consequences for urban economic change. The ESRC programme was concerned with the distributive effects of urban economic change, and particularly the employment effects on disadvantaged urban residents of employment decline and growth. Much attention has been paid to the plight of the disadvantaged in areas of declining employment. But more attention must be directed to how the disadvantaged fare in expanding labour markets if we are to identify clearly the factors influencing the allocation of employment among groups and to design more effective policies for those whose job prospects are least promising.

The ESRC programme was also distinguished by its concern for the effects of public policies on urban economic change, and the identification of those policy factors which impede and facilitate the adjustment of urban economies and the improved distribution of employment opportunities. The concern with public policy has extended beyond area development and employment creation policies to address other central government policies relevant to urban development and employment. This public policy theme is advanced in this volume through studies of national manpower and housing policies.

In recent years manpower programmes have expanded substantially as a component of central government policy, and now constitute a major expenditure on employment-related problems at the local level. The expansion has been linked with a concern for the human resources dimension of national economic growth and the problems of unemployment. But how relevant are national manpower programmes to urban economic development and labour-market problems? Are there important constraints on the effectiveness of these programmes in the urban setting and on their support for the development of innovative local manpower programmes? What are the requirements for effective urban manpower policies? These issues are addressed in an essay on manpower policies and urban employment and training initiatives.

The last essay in this volume examines the effects of housing policy on the inner city problems of concentrated poverty and unemployment and disinvestment by people and firms. How do housing policies affect urban economic development and employment creation, the geographic concentration of social and economic problems, and the mobility of urban residents? How can housing policy be made more supportive of urban economic policy objectives?

Harold Wolman's comparative analysis of urban economic performance is an initial useful contribution to putting British urban economic experience in perspective and offers some important findings on the factors influencing urban economic performance by extending to the international level the ESRC programme's approach of analysing diversity in local economic experience. It also illuminates the structural and spatial aspects of economic change in Western advanced industrial nations during the decade of the 1970s.

Unemployment rose relative to the respective national rates in the urban systems of all four countries (the USA, Britain, West Germany, and the Netherlands) studied by Wolman during the economic difficulties and transformations of the 1970s. The performances of Britain and the Netherlands were the worst. However, urban areas continue to play a dynamic economic role. The urban systems of all three European nations continued to outperform their national economies in terms of employment growth. Interestingly, the USA experienced the greatest diversity in economic performance (that is, increases in employment growth and unemployment rates) among its cities, while variations in employment growth were much greater among urban areas in Britain than in Germany. German cities have been generally stronger economically despite rising unemployment problems.

Wolman's analysis underscores the important regional dimension to urban economic decline across advanced industrial nations. Economic growth is occurring away from the older industrial regions, as evidenced by the heavy concentration of better-performing cities in the South and East of Britain, the West of the USA, and the South of Germany. Only one of Britain's poor urban economic performers (Medway Towns) was located in the economically more successful regions of the country. But, in contrast to the USA, the relative economic problems of the European urban areas studied persisted in the last decade. The relative economic performance of those European urban areas which began the 1970s with the highest unemployment rates worsened during the decade.

The development of the manufacturing sector as an element of urban economic regeneration strategies, as well as of national economic development policy, is an important issue surrounded by great dispute. It is Wolman's findings regarding the importance of the inherited industrial structure, and particularly the influence of the manufacturing sector on urban economic performance, which stand out most. With the interesting exception of Germany, those urban areas whose industrial structures had been dominated by manufacturing employment experienced poor economic performance. However, the manufacturing sector was not *necessarily* an impediment to successful urban economic performance, for in many of the successful urban performers, particularly in the USA and Germany, manufacturing employment grew rapidly. Indeed, in Britain the difference between areas with successfully performing urban economies and those areas with unsuccessful urban economies was related much more to differences in growth in the manufacturing sector than to differences in growth in the service sector, and the performance of the manufacturing sector had a stronger influence than that of the service sector on changes in both urban employment and the unemployment rate. However, in Britain, as in the USA, the manufacturing sector performed less well where manufacturing had dominated the urban economy— that is, in the older industrial centres. This can be seen in the contrast between the performance of the manufacturing sectors of Bristol and those of Glasgow, Newcastle, and Birmingham.

The German experience is also significant. A preponderance of manufacturing employment did not result in poorer urban economic performance, or poorer urban manufacturing performance. This suggests the need to look in more depth at the character of the urban manufacturing sector in assessing its role in particular urban economies and its potential for contributing to

local economic growth. The quality of the nation's manufacturing sector—
and here there is a clear disparity between Germany and Britain—is also of
prime importance. The success of more competitive, advanced manufacturing sectors can contribute significantly to urban economic performance.

The Wolman essay cites the work of Noyelle and Stanback (1984) in the
USA: this proposes that it is changes in the functions of urban economies in
this period of major structural economic change which have had the significant influence on urban economic performance and help to explain the variations in urban performance. Noyelle and Stanback conclude that in the
USA during the 1970s urban areas specializing in manufacturing production, and manufacturing production with corporate headquarters functions,
performed poorly while areas specializing in consumer-oriented services, in
particular, and in government and education, the defence industry, and
diversified producer services performed substantially better than the nation.
The analysis of Begg and Moore in Chapter 3 explores the relevance of urban
structural and functional change to variations in urban economic performance in Britain.

Their study of the fastest and slowest growing British cities between 1971
and 1981 again demonstrates the redistribution of economic growth from the
industrial North towards smaller cities in the South of the country. Structur-
ally, the fastest growing cities showed greater specialization in advanced
manufacturing, and a greater concentration in rapidly growing service sectors, particularly producer services. The slowest growing cities had a greater
concentration of employment in traditional manufacturing industries and
standard manufactures open to heavy international competition, and a
below-average share in labour-intensive and more rapidly growing service
sectors. The study indicates the very rapid growth rate in the 1970s of
advanced producer services. This sector, with only a small proportion of the
nation's employment, created the same number of jobs as did the very large
public services sector. The analysis also confirms that urban economic performance is significantly influenced by local characteristics (for example, city
size, New Towns, peripherality) and cannot be explained simply by an area's
share of nationally declining industries. Thus, both the Wolman study and
that by Begg and Moore indicate that the search for explanations of diversity
in urban economic performance requires careful assessment of the effects of
area characteristics, including the distinctive features of the local industrial
structure, on local economic change.

Among the policies which have been proposed to address the disproportionate incidence of unemployment among inner city residents are local
employment creation and facilitation of the movement of disadvantaged residents out of areas of economic decline. However, the ESRC programme's
research has raised serious questions about the potential effectiveness of such
policies. It suggests that more significant than economic decline in explaining the higher levels of inner city unemployment are certain personal charac-

teristics—for example, race, skill, and age—of inner city residents which disadvantage them in the competition for employment. Thus, inner city employment opportunities are diffused throughout the urban labour market to the benefit of others viewed as more qualified. Moreover, inequalities in employment opportunities persist in economically healthier areas such as Bristol. Buck and Gordon address these vital labour-market issues through an analysis of the employment experiences of disadvantaged groups in expanding labour markets as contrasted with areas of economic decline.

Are those disadvantaged in the labour market better off in employment terms in areas of employment growth? The answer is yes. While the disparities between the disadvantaged and other groups persist in growth areas, their risks of unemployment and longer durations of unemployment are reduced, relative earnings levels are improved, and both the movement out of unskilled jobs and the proportion of workers in stable employment are increased. Moreover, policies to assist the disadvantaged, such as training, appear to be more successful in growth areas. Thus, there is value in promoting the out-migration of inner city residents in the southern growth regions of the country, both to improve the circumstances of the out-migrants and to reduce the concentrations of inner city deprivation. However, the feasibility of such dispersal policies would depend on the removal of housing and planning impediments to the freer movement of the poor. Moreover, without national growth in employment, migration policies will only redistribute the unemployment problem among areas.

In contrast, employment disparities between groups are increased in areas of economic decline. The disadvantaged are bearing the brunt of urban economic adjustment. And local employment growth in the context of declining economic regions, such as the situation of New Towns in the North, has little effect on the unemployment of the disadvantaged because of the demands for employment in the wider region and the diffusion of jobs to those more competitive in the labour market. It is the growth areas concentrated within the generally successful economic environment of southern England which deliver the benefits of growth to the disadvantaged. Therefore, effective regional development and employment policies are essential to addressing the employment problems of distressed northern cities. Within the cities, employment policies will need to be directly targeted on disadvantaged groups, and not simply spatially targeted, if they are to be of benefit. Job preservation policies which reduce the numbers of disadvantaged urban residents falling into unemployment would also be beneficial.

Central government policies which are vital to an effective response to urban labour-market and economic development problems need to be mobilized in support of urban policy objectives. In order to achieve that objective, government policies need to be assessed in terms of their current responsiveness to urban economic needs and to identify any impediments to positive government participation in local development efforts. McArthur and McGregor contribute to this end with their study of the Manpower Services Commission (MSC) and innovative urban employment and training policies. They highlight a number of examples of local innovative manpower policies but also the limited support provided by current national policies for needed innovations.

Their study indicates a number of manpower policy requirements for improving urban employment and development policies:

(*a*) Making use of training—this is, investment in human capital—as an element of urban economic development strategies, and not simply social policy

(*b*) Giving adequate and effective attention to the employment and training needs of the long-term unemployed and disadvantaged

(*c*) Allowing national manpower programmes to be tailored to the specific needs of particular distressed urban economies and to local economic policy objectives

(*d*) Providing the organizational framework for effective local delivery of manpower policies

(*e*) Providing a role for local authorities in national manpower policies which would facilitate the linkage of urban economic development and manpower policies.

The research identifies a number of impediments in current manpower policies. The increasing emphasis on employer-based and market-oriented employment and training schemes limits the resources which are directed to distressed urban economies with inadequate private sector demand, and the training given to disadvantaged clients. Cost-effectiveness criteria reinforce an aversion to addressing the costlier needs of disadvantaged clients and areas. Employment programmes give insufficient attention to the longer-term training and employment requirements of the disadvantaged, to addressing their special handicaps, and to the need for innovative employment creation measures. The authors recognize the important link between economic and social objectives when they note the negative effects on an area's potential for economic growth of the declining employability of a large concentration of long-term unemployed.

Training has not been accepted by central government as an element of urban economic development policies as it has with regard to national economic development. Training programmes are necessary to strengthen the human resources of distressed cities in order to increase their potential for future economic growth, and should not be constrained by the lack of current business demand. This is especially important in a period in which the knowledge and skills of the labour force are increasingly important to local economic performance. Local authority efforts to develop manpower policies related to local needs or longer-term strategic objectives have not been encouraged by the MSC. National policies must be more flexible to allow for decentralized administration, local tailoring, innovation, and co-ordination of programmes, and co-operation with area development policies, based on adequate labour-market analysis.

The Ermisch and Maclennan essay identifies a similar, destructive separation between housing and urban policy, and then proceeds to examine the influences of housing policy on urban job creation, social mix and concentrations of the poor in inner cities, and the mobility of inner city residents. In the past, the combination of the private housing market and council housing policy contributed to the decentralization of more affluent home-owners, social polarization, and the concentration of the disadvantaged in inner city areas. More recently, government housing policy has had conflicting effects on the problem of concentrated inner city deprivation. Housing rehabilitation policies and subsidies to encourage home-ownership combined with increasing investment by building societies have had positive effects on older urban neighbourhoods with spillover effects on surrounding neighbourhoods. These urban housing improvements have contributed to a reduction in social concentration by retaining and attracting households with rising incomes. Against this, substantial reductions in government support for council housing, rising local authority rents, and sales of council housing have increased the concentration of the lowest income groups in council housing and exacerbated the problems of declining inner area neighbourhoods.

The authors argue that appropriate housing and environmental improvement policies could lead the way towards urban economic regeneration. Rehabilitation would promote greater inner area social diversity and private investment while providing construction and service sector employment to less skilled inner city residents. Improving the council housing sector would also retain socio-economic diversity and provide employment opportunities. Diminishing social polarization is more likely through an improved mix of inner city owner-occupied and rental housing than by breaching the serious impediments to out-migration of the lowest income groups. Public policy should seek to combine inner area housing improvement policies with facilitation of the out-migration of the lowest-income inner city residents in the prosperous southern region. Thus, the authors are concerned by government's retreat from its earlier support for inner urban area housing improvements.

The studies in this volume emphasize once again the dual but linked issues of promoting the long-term economic restructuring and development of the urban regions of the country, particularly in the North, and ameliorating employment disadvantages and reducing concentrated deprivation throughout urban areas. They explore the policy implications of structural change and labour-market processes, and underline the need for policies to focus clearly and directly on removing impediments to adjustment and facilitating

Critical Issues II

concerted efforts by diverse government policies to address urban economic concerns.

Reference

Noyelle, T. J. and Stanback, T. M., Jr. (1984), *The Economic Transformation of American Cities*, Totowa: Rowan and Allanheld.

2

Urban Economic Performance:
A Comparative Analysis

Harold Wolman

Introduction

The economic problems experienced by large urban areas in Britain, while severe, are by no means unique. Many urban areas in other advanced Western nations have faced similar economic difficulties. Thus, while the Economic and Social Research Council's (ESRC's) Inner Cities Research Programme was broadly designed to explore urban economic change and its consequences in Britain and to examine competing hypotheses purporting to explain the performance of Britain's urban economies, the objective of the research reported in this chapter is to place the British experience in a comparative perspective. By this means it is hoped that additional light can be thrown on British urban economic performance and on the validity of some of the hypotheses set forth to explain it.

This objective is pursued primarily through examination of variations in the performance of urban economies during the period 1970–80 within each of several countries. The primary questions to be answered are why the economic performance of some urban areas has been more successful than that of others and whether there are generalizations holding across countries that can be drawn in response to this question. The research attempts to explain whether variation in performance is related to a variety of competing hypotheses which have been offered, within the ESRC Programme and elsewhere, as possible causes of urban economic performance. These include regional economic performance, urban area size, urban economic function, industrial structure, sectoral performance, labour-force quality, wage rates, and amenities. In some cases, data for examining these hypotheses are not available for all of the countries in our study or are available for only one country (usually the USA). We none the less examine the data that are available. However, a variety of other plausible hypotheses are not examined at all owing to a lack of data.

This work complements existing cross-national studies (see, in particular, Hall and Hay 1980, and van den Berg *et al.* 1982) which are concerned primarily with population change and with defining stages of urbanization rather than with economic performance. For a comprehensive review of both the comparative research and single-country literature on 'urban decline' see Cheshire *et al.* (1986).[1]

Research Design and Methodology

The study examined all urban areas of more than 200 000 in population in Great Britain,[2] the USA, West Germany, and the Netherlands. Urban area was defined conceptually not in terms of political jurisdiction or boundaries, but as the functional economic area, constituting an integrated labour market, surrounding an urban core. However, the exact definition of this geographic entity differed from country to country (see Appendix 1).

Within each country a set of good urban economic performers and a set of poor urban economic performers (in relative terms) were identified. Guided by the hypotheses suggested in existing literature, the analysis then attempted to determine in what ways the good performers differed from the poor performers.

Obviously the definition of economic performance is critical. There is no universally accepted definition of what constitutes good economic performance for an urban area. A variety of measures have been used including, *inter alia*, the level and rate of change of total employment, unemployment rate, value added or gross local product, and per capita earnings from employment. All of these measures clearly capture some aspect of economic performance; none captures the entire concept.

Our objective is not to engage in a once-and-for-all effort to posit the best definition for economic performance, but rather to select a measure or measures which capture important aspects of the concept and correspond to—or at least do not violate—common sense. In addition, the measures must be ones for which data are realistically available for cross-national studies using the urban area as the unit of analysis. For these purposes we utilize two measures: (1) change in an area's total employment, and (2) change in an area's unemployment rate.

The first, change in an area's total employment, measures the ability of the area's economy to generate jobs. The second, change in an area's unemployment rate, measures the ability of the area's economy to clear labour markets. In many studies—and indeed in much policy discussion—employment growth (or decline) is taken as the sole measure of an area's economic performance. However, this violates our common-sense criterion. Most people would not consider an area in which employment had increased by 5 per cent but the unemployment rate had increased from 6 to 12 per cent to have performed better than an area in which employment had not increased at all but the unemployment rate had fallen from 6 to 3 per cent. The two measures, taken in conjunction with one another, provide a clearer picture of performance.

It must be stressed that economic performance is an aggregate concept, characterizing the operations of the urban economy as a whole. Economic performance, as thus defined, is not concerned with either the level or the distribution of economic welfare, which relates to the well-being of individuals or groups rather than economies. There is no *necessary* connection between growth of employment in an area and an increase in economic well-being. At first glance there would appear to be a direct relationship between a reduction in the unemployment rate and an increase in economic well-being, since the possession of a job, in itself, may be seen as a good. However, if we are concerned with the well-being of those in the community, as it exists at the beginning of the period, then change in the unemployment rate over time, as conventionally measured, cannot give direct evidence on change in well-being. This is because the nature of the community will have changed as a consequence of migration and the nature of the labour force will have changed as a consequence of shifts in activity rates. Thus, for example, a reduction in unemployment among those who lived in the community at point A may be obscured by an increase in the unemployment rate caused by in-migration of unemployed to the area between point A and point B.

While the unemployment rate does provide information on changes in the ability of an area's economy to clear labour markets (as they are constituted at any point in time), it thus does not provide reliable information on changes in economic well-being. The relationship between economic performance and economic well-being is an important empirical question, but one this research deals with only peripherally.

It is also important to note that the performance measures, as used in this study, relate not to levels, but to changes over time. Thus an urban area with a 6 per cent rate of unemployment in 1970 but a 9 per cent rate in 1980 performed better during the 1970–80 period than an area with a 4 per cent rate of unemployment in 1970 and an 8 per cent rate in 1980. (Obviously this is not meant to imply that the area with higher unemployment is 'better off' than that with the lower rate, but that over a defined period of time its performance was better.) The time periods analysed are 1970–80 or the closest dates to that period (1971–81 for Britain). However, satisfactory German data were only available for the 1978–83 period. Thus differences between Germany and the other countries could relate to the time period studied.

As the above implies, the performance of an area is a term which is meaningful only relative to that of other areas; accordingly, we compare the performance of an urban area primarily to the performance of the entire national economy, or we compare 'good performers' to 'poor performers'. In general, 'good performers' are those urban areas which perform better than the national average in terms of *both* unemployment rate change (lower) and total employment change (higher), while 'poor performers' are those areas which perform worse than the national average on both indicators.[3] Good performers and poor performers thus together constitute less than the total number of urban areas in the analysis. The remainder of 'mixed performers'

are left out of the analysis so that the contrast between good and poor performers can be observed; however, the mixed performers are included in the
regression analyses. Of the five British urban areas selected for case-studies
in the ESRC's Inner Cities Research Programme, Bristol fell into the class of
good performers, Birmingham and Newcastle were poor performers, and
London and Glasgow were mixed performers.[4]

The good and poor performers in each of the countries are listed in
Table 2.1.

Relationship of Performance Variables to Each Other

We begin by examining how the performance variables were related to each
other. Clearly there is some interaction between the measures. Thus an
urban area growing rapidly in employment might attract in large numbers of
mobile unemployed workers and/or experience a rise in activity rates which
would result in an increase in unemployment rate simultaneously with
expanding employment. How frequently did these ambiguous cases occur?

In general, high employment growth (employment growth above the
national average) was associated with a low increase in unemployment rate
(an increase in the unemployment rate less than the national average). This
was particularly pronounced in the USA, where 90 per cent of the urban
areas with above-average employment growth ($n = 52$) also had below-
average increases in unemployment, compared to 75 per cent in Britain ($n =
32$) and 73 per cent in Germany ($n = 37$). Similarly, low employment growth
was associated with a high increase in unemployment, particularly in Britain
(90 per cent of areas with below-average employment growth had an above-
average increase in the unemployment rate) and Germany (83 per cent). In
the USA, however, the comparable figure was only 58 per cent. Thus, in the
USA, while high employment growth virtually assured a good unemployment performance, slow economic growth did not inevitably condemn an
area to substantial increase in the unemployment rate (see Table 2.2).

It seems clear from these results that, while an area growing in employment may induce gains in its labour force, the net result in the vast majority
of cases is still likely to be lower unemployment rates than would otherwise
have been the case. Correlation analysis supports the close inverse relationship between the rate of change in urban area employment and changes in
the unemployment rate. In Britain, the correlation (r) between these two variables was -0.68; in the USA it was -0.55, and in Germany -0.56.

Performance over Time

To what extent did urban areas with the worst economic conditions at the
beginning of the study experience relatively better performance and thus
improve their relative standing over the course of the period? In Great

Table 2.1. *The Good and Poor Urban Performers considered in the Comparative Analysis*

Country	Good performers	Poor performers
Britain	Aberdeen	Ashton and Hyde
	Aldershot–Farnborough	Birkenhead and Wallasey
	Blackpool	Birmingham
	Bournemouth	Bolton
	Brighton	Bradford
	Bristol	Cardiff
	Cambridge	Coventry
	Derby	Dundee
	Edinburgh	Huddersfield
	Exeter	Hull
	Guildford	Liverpool
	High Wycombe	Medway Towns
	Ipswich	Middlesbrough
	Northampton	Motherwell
	Norwich	Newcastle
	Nottingham	Oldham
	Oxford	Pontypridd
	Reading	Stoke-on-Trent
	Southampton	Sunderland
	York	Walsall
		Wigan
		Wolverhampton
USA	Albuquerque, N. Mex.	Akron, Ohio
	Anaheim–Santa Anna–Garden Grove, Calif.	Allentown–Bethlehem–Easton, Pa.–NJ
	Austin, Tex.	Baltimore, Md.
	Bakersfield, Calif.	Buffalo, NY
	Colorado Springs, Colo.	Canton, Ohio
	Dallas–Fort Worth, Tex.	Chattanooga, Tenn.–Ga.
	Denver–Boulder, Colo.	Chicago, Ill.
	Eugene–Springfield, Oreg.	Cleveland, Ohio
	Fresno, Calif.	Dayton, Ohio
	Fort Lauderdale–Hollywood, Fla.	Detroit, Mich.
	Houston, Tex.	Erie, Pa.
	Jackson, Miss.	Flint, Mich.
	Las Vegas, Nev.	Fort Wayne, Ind.
	Lexington–Fayette, Ky.	Gary–Hammond–E. Chicago,Ind.
	Orlando, Fla.	Hamilton–Middletown, Ohio
	Oxnard–Ventura, Calif.	Harrisburg, Pa.
	Phoenix, Ariz.	Huntington–Ashland, W. Va.– Ky.–Ohio
	Portland, Oreg.–Wash.	Huntsville, Ala.
	Raleigh–Durham, NC	Jersey City, NJ
	Sacramento, Calif.	Johnstown, Pa.
	Salt Lake City–Ogden, Utah	Lima, Ohio
	San Diego, Calif.'	Lorain–Elyria, Ohio
	San Jose, Calif.	Louisville, Ky.–Ind.
	Santa Barbara, Calif.	

Table 2.1.—*cont.*

Country	Good performers	Poor performers
	Santa Rosa, Calif.	Newark, NJ
	Tampa–St. Petersburg, Fla.	New York, NY
	Tucson, Arizona	North–east Pennsylvania
	Tulsa, Okla.	Paterson–Clifton–Passaic, NJ
	Vallejo–Fairfield–Napa, Calif.	Peoria, Ill.
	Wichita, Kan.	Providence–Warwick–Pawtucket, RI–Mass.
		Reading, Pa.
		Rockford, Ill.
		Saginaw, Mich.
		South Bend, Ind.
		Toledo, Ohio–Mich.
		Wilmington, Del.–NJ.–Md.
		York, PA.
		Youngstown–Warren, Ohio
Germany	Amberg	Bielefeld
	Ansbach	Bochum
	Aschaffenburg	Bremen
	Augsburg	Bremerhaven
	Baden-Baden	Cologne
	Bonn	Dortmund
	Freiburg	Duisburg
	Heilbronn	Düsseldorf
	Ingolstadt	Essen
	Kempten	Hagen
	Landshut	Hanover
	Memmingen	Hamburg
	Munich	Kassel
	Regensburg	Kiel
	Rosenheim	Krefeld
	Stuttgart	Lübeck
	Speyer	Mönchen-Gladbach
	Würzburg	Saarbrücken
	Zweibrücken	Wilhelmshaven
		Wuppertal
Netherlands	Enschede–Hengelo	Eindhoven
	Haarlem	Groningen
	The Hague	Haarlem—Kerkrade
	Tilburg	Nijmegen
	Utrecht	Rotterdam

Britain, Germany, and the Netherlands, those urban areas with the highest unemployment rates at the beginning of the period tended to have the greatest increase in unemployment rates over the period: the worst became relatively worse. Thus the correlation (r) between the base-year unemployment rate and percentage point increase in unemployment over the period covered

Table 2.2. *Employment Growth and Unemployment Performance in Britain, the USA and Germany*

Country	Total employment growth performance	Unemployment performance	
		Good	Poor
Britain	Good	24	8
	Poor	3	26
USA	Good	47	5
	Poor	40	57
Germany	Good	27	10
	Poor	4	20

was 0.60 for Germany, 0.38 for the Netherlands, and 0.58 for Britain. In Britain, each additional one percentage point of unemployment in 1971 was associated with, on average, an additional increase of 0.75 percentage points in the unemployment rate between 1971 and 1981. In the USA, however, there was a modest negative correlation ($r = -0.26$): the worst became relatively better. Each additional percentage point of 1970 unemployment was associated with a 0.40 percentage point lower increase in the unemployment rate between 1970 and 1980.

Explaining Performance

What factors accounted for the relative economic success of some urban areas and the relatively poor performance of others?[5] We begin this analysis by examining how the two groups differed from each other.

Region

It has frequently been asserted that the performance of urban economies is dependent largely on the performance of the economies of the region of which they are a part. Our analysis, indeed, showed a very strong relationship between regional and urban economic performance. In all three of the countries with well-developed urban systems (the USA, Britain and Germany), the division between good and poor urban performers was primarily a regional division (see Table 2.3).

In Britain, urban areas in regions with strong economies (the South and East) performed better over the course of the decade than did urban areas in the rest of the country. The three regions with good economic performances (employment growth above the national average and change in unemployment rate below the national average) were the South West, East Anglia, and the East Midlands. The South East region had the second-lowest change in unemployment rate (6.1 per cent), well below the national average (7.9 per cent), but a slightly below-average rate of employment growth. Of the twenty

Table 2.3. *Urban Economic Performance and Region in Britain, the USA, and Germany*

	Good performance[a]			Poor performance[a]		
	Observed frequency (O)	Expected frequency[b](E)	Ratio O : E	Observed frequency (O)	Expected Frequency[b](E)	Ratio O : E
Britain:						
Southern and Eastern Regions	16	(7.5)	2.13	1	(8.3)	0.12
Remainder of country	4	(12.5)	0.32	21	(13.7)	1.53
USA:						
South and West	29	(13.6)	2.13	3	(16.3)	0.18
Northeast and Midwest	1	(16.4)	0.06	33	(19.7)	1.68
Germany:						
2 southern lander	16	(7.5)	2.13	0	(7.9)	0
Rest of country	3	(11.5)	0.26	20	(12.1)	1.65

[a] See Appendix 2 for definitions of good and poor performers.
[b] Expected frequency calculated as

$$\frac{\text{All urban areas over 200 000 in region}}{\text{All urban areas over 200 000 in nation}} \times \text{No. of good performers in nation}$$

(for expected frequency of good performers) or

$$\frac{\text{All urban areas over 200 000 in region}}{\text{All urban areas over 200 000 in nation}} \times \text{No. of poor performers in nation}$$

(for expected frequency of poor performers).

good overall urban economic performers, sixteen were in these four regions, while only four (Aberdeen, Edinburgh, Blackpool, and York) were in the remainder of the country. Of the 'poor' performers, only one (Medway Towns) was located in these four high-performing regions. The four southern regions had more than twice the number of good performers and only one-eighth the number of poor performers as would have been expected had good and poor performers simply been distributed proportionately by region.

In the USA, the regional division was even more pronounced. Of the thirty 'good-performing' urban areas between 1970 and 1980, twenty-nine were located in the southern, southwestern, or western regions of the country—the economically thriving sunbelt—while only one (Wichita) was located in the Northeast or Midwestern regions. However, of the thirty-six poor performers, thirty-three were located in the Northeastern or Midwestern regions. The Northeast and Midwest had 1.68 times the number of poor performers and less than 1/15 the number of good performers as would have been expected, while the South and West had one-fifth the number of poor performers and more than twice the number of good performers as would have been expected.

In Germany, sixteen of the nineteen good urban performers were located in the two southern lander of Baden-Württemburg and Bavaria, two in the Rhineland, and one in North Rhine-Westphalia; none were located in the two northernmost lander of Schleswig-Holstein and Lower Saxony. However, seven of the twenty poor performers were located in the northernmost regions (including Hamburg and Bremen), eleven in North Rhine-Westphalia, and none in the two southernmost lander.

Urban area size

A variety of recent studies have found a relationship between urban area population size and manufacturing performance. For example, Fothergill and Gudgin conclude (1982, 8):[6]

Britain's cities are experiencing a rapid loss of manufacturing jobs while small towns and rural areas are quite successful in retaining and expanding their manufacturing base. As a general rule, the larger and more industrial a settlement the faster its decline.

Does this same pattern hold true in terms of our measures of overall urban economic performance? It is true that in all three countries the very largest areas tended to be greatly over-represented in the 'poor performers'. In Germany, 45 per cent of the poor-performing urban areas were over 1 million in population, more than 1.5 times more than would have been expected if performance did not vary at all by population size, while only 10.5 per cent of the good performers were above 1 million, nearly three times fewer than would have been expected (see Table 2.4).

Table 2.4. *Urban Economic Performance and Population Size in Britain, the USA, and Germany*

Population size (000s)	Good performers				Poor performers			
	No. observed	% observed (O)	% expected (E)	Ratio O : E	No. observed	% observed (O)	% expected (E)	Ratio O : E
Britain								
200–300	10	50	49.2	1.02	10	45.5	49.2	0.92
300–500	7	35	27.9	1.25	7	31.8	27.9	1.14
500–1000	3	15	14.8	1.01	3	13.6	14.8	0.92
1000–2000	0	0	6.6	0	2	9.1	6.6	1.38
Over 2000	0	0	1.6	0	0	0	1.6	0
Total	20	100			22	100		
Germany								
200–300	0	0	1.6	0	0	0	1.6	0
300–500	8	42.1	31.2	1.35	4	20	31.2	0.64
500–1000	9	47.4	37.7	1.26	7	35	37.7	0.93
1000–2000	0	0	19.7	0	6	30	19.7	1.52
Over 2000	2	10.5	9.8	1.07	3	15	9.8	1.53
(Over 1000)	(2)	(10.5)	(29.5)	(0.36)	(9)	(45)	(29.5)	(1.52)
Total	19	100			20	100		
USA								
200–300	8	26.7	27.1	0.99	11	29.7	27.1	1.10
300–500	9	30.0	25.8	1.16	8	21.6	25.8	0.84
500–1000	5	16.7	25.1	0.67	11	29.7	25.1	1.18
1000–2000	7	23.3	12.3	1.89	1	2.7	12.3	0.22
Over 2000	1	3.3	9.7	0.34	6	16.2	9.7	1.67
(Over 1000)	(8)	(26.6)	(22.0)	(1.21)	(7)	(18.9)	(22.0)	(0.86)
Total	30	100			37	100		
USA: Northeast and Midwest only								
200–300	5	25	26.5	0.94	5	23.8	26.5	0.90
300–500	5	25	23.5	1.06	2	9.5	23.5	0.40
500–1000	3	15	25.0	0.60	9	42.9	25.0	1.72
1000–2000	6	30	10.3	2.92	1	4.8	10.3	0.47
Over 2000	1	5	14.7	0.34	4	19.0	14.7	1.29
(Over 1000)	(7)	(35)	(25.0)	(1.40)	(5)	(23.8)	(25.0)	(0.95)
Total	20	100			21	100		

In the USA, six of the thirty-seven poor-performing urban areas had populations in excess of 2 million, 1.67 times the number that would be expected if performance was unrelated to size, while only one area over 2 million in population was a 'good-performer', three times less than would have been expected. (However, in the 1–2 million size category, the situation was reversed. There were seven good performers compared to only one poor performer.) Controlling for region made little difference in the USA; roughly the same pattern held in the Northeast–Midwestern states.

In Britain, two areas of over 1 million were poor performers, 1.38 times more than would have been expected if performance had been distributed without regard to size, while none were good performers.

However, in the USA and Great Britain, except for the very largest areas, there was little evidence of a size *gradient*, in which the smaller the area, the better the economic performance. Thus, in the smallest population size group (200 000–300 000 in population), the percentage of good performers in the USA was exactly what would have been expected if performance did not vary by size group, and the percentage of poor performance was actually slightly higher than would have been expected. In the Northeast–Midwest region, both good and poor performers with 200 000–300 000 population had slightly less than expected frequencies. In Britain, the percentage of good performers in this size group was nearly exactly as expected, while the percentage of poor performers was slightly lower than expected.

However, in both Britain and the USA, the next-largest size group (300 000–500 000) had more good performers than expected (1.25 and 1.16 times respectively). This size group in Britain also had more poor performers (1.14 times more than expected), while it had fewer than expected in the USA (0.84 times what was expected). The gradient effect appeared to exist only in Germany where smaller-size urban areas performed consistently better than larger-size ones through all the size categories.

Thus, except for the very largest urban areas, the two groups of good and poor performers did not appear to differ markedly with respect to size in either the USA or Britain. However, if we examine the two performance variables—change in total employment and change in unemployment rate—separately, does a different picture emerge? Here we turn for analysis to our entire population (all urban areas over 200 000) rather than solely to the groups of good and poor performers.

In both Britain and Germany, a slight size gradient did appear to exist with respect to employment growth. Smaller areas were more likely to have grown at a rate above the national average than were larger areas, and the smaller the size category the greater the ratio of above-average growers to the number which would have been expected if size had no effect. In the USA, however, no such gradient appeared: indeed urban areas with populations between 1 and 2 million were over-represented among above-average growers and under-represented among below-average growers[7] (see Table 2.5).

The relationship between population size and unemployment rate change was even weaker. As in overall economic performance, the largest urban areas in all three countries appeared over-represented among those areas experiencing above-average increases in unemployment rates and under-represented among areas with below-average increases. Looking at all size categories, a modest gradient appeared to exist in Germany with smaller areas performing better than larger areas, but a reverse gradient seemed to exist in the USA, a pattern which persisted even when region was controlled for. No evidence of a size gradient appeared in Britain; indeed areas in the 500 000–1 000 000 size category were over-represented among areas with below-average increases in unemployment rates by a factor of 1.25 and under-represented among areas with above-average increases by a factor of 0.80.

Table 2.5. *Employment Growth and Unemployment Rate Performance by Size of Urban Area in Britain, the USA, and Germany*

Population size (000s)	Employment growth							
	Above-average growth				Below-average growth			
	No. observed	% observed (O)	% expected (E)	Ratio O : E	No. observed	% observed (O)	% expected (E)	Ratio O : E
Britain								
200–300	19	54.5	49.2	1.11	11	39.3	49.2	0.80
300–500	10	30.3	27.9	1.09	7	25.0	27.9	0.90
500–1000	4	12.1	14.8	0.82	5	17.9	14.8	1.21
1000–2000	0	0	6.6	0	5	17.9	6.6	2.71
Over 2000	0	0	1.6	0	0	0	1.6	0
(Over 1000)	(0)	(0)	(8.2)	0	(5)	(17.9)	(8.2)	(2.18)
Total	33	100	100		28	100	100	
Germany								
200–300	0	0	1.6	0	1	4.1	1.6	2.56
300–500	14	37.8	31.2	1.21	4	16.7	31.2	0.53
500–1000	16	43.2	37.7	1.15	8	33.3	37.7	0.88
1000–2000	5	13.5	19.7	0.69	7	29.2	19.7	1.48
Over 2000	2	5.4	9.8	0.55	4	16.6	9.8	1.69
(Over 1000)	(7)	(18.9)	(29.5)	(0.64)	(11)	(45.8)	(29.5)	(1.55)
Total	37	100	100		24	100	100	
USA								
200–300	15	28.3	27.1	1.04	30	30.3	27.1	1.12
300–500	16	30.2	25.8	1.17	24	24.2	25.8	0.94
500–1000	11	20.8	25.1	0.83	22	22.2	25.1	0.88
1000–2000	10	23.2	12.3	1.89	9	9.1	12.3	0.74
Over 2000	1	1.9	9.7	0.20	14	14.1	9.7	1.45
(Over 1000)	(11)	(20.8)	(22.0)	(0.95)	(23)	(23.2)	(22.0)	(1.05)
Total	53	100	100		99	100	100	
USA: South and West only								
200–300	13	26.5	30.5	0.87	12	36.4	30.5	1.19
300–500	15	30.6	26.8	1.14	9	27.3	26.8	1.02
500–1000	10	20.4	20.7	0.99	6	18.2	20.7	0.88
1000–2000	10	20.4	15.9	1.28	2	6.1	15.9	0.38
Over 2000	1	2.0	6.1	0.33	4	12.1	6.1	1.98
(Over 1000)	(11)	(22.4)	(22.0)	(1.02)	(6)	(18.2)	(22.0)	(0.83)
Total	49	100	100		33	100	100	

Urban function

Is economic performance related to the function urban areas perform within the urban system? Noyelle and Stanback (1984) argue that widespread economic transformations—in particular, the decline in importance and decentralization of production employment and the increase in importance of producer services—have brought about major changes in the structure of the US urban system and in the functions that various urban areas perform. They identify four major functional types, some of which, in turn, are further subcategorized:

(1) *Diversified service centres* are 'centers specializing in the provision of intermediate service activities such as headquarters functions, producer services, distributive services, and non-profit and government activities': these are subclassified into national, regional, and subregional centres

Table 2.5. *Employment Growth and Unemployment Rate Performance by Size of Urban Area in Britain, the USA, and Germany—cont.*

Population size (000s)	Unemployment rate							
	Below-average growth				Above-average growth			
	No. observed	% observed (O)	% expected (E)	Ratio O : E	No. observed	% observed (O)	% expected (E)	Ratio O : E
Britain								
200–300	14	51.9	49.2	1.05	16	47.1	49.2	0.96
300–500	7	25.9	27.9	0.98	10	29.4	27.9	1.05
500–1000	5	18.5	14.8	1.25	4	11.8	14.8	0.80
1000–2000	1	3.7	6.6	0.56	4	11.8	6.6	1.79
Over 2000	0	0	1.6	0	0	0	1.6	0
(Over 1000)	(1)	(3.7)	(8.2)	(0.45)	(4)	(11.8)	(8.2)	(1.43)
Total	27	100	100		34	100	100	
Germany								
200–300	0	0	1.6	0	1	3.1	1.6	2.56
300–500	11	37.9	31.2	1.21	7	21.9	31.2	0.70
500–1000	13	44.8	37.7	1.19	11	34.4	37.7	0.91
1000–2000	2	6.9	19.7	0.35	10	31.3	19.7	1.59
Over 2000	3	10.3	9.8	1.05	3	9.4	9.8	0.96
(Over 1000)	(5)	(17.2)	(29.5)	(0.58)	(13)	(40.7)	(29.5)	(1.38)
Total	29	100	100		32	100	100	
USA								
200–300	23	26.1	27.1	0.96	19	29.7	27.1	1.10
300–500	25	28.4	25.8	1.10	16	25.0	25.8	0.97
500–1000	20	22.7	25.1	0.90	15	23.4	25.1	0.93
1000–2000	15	17.0	12.3	1.38	4	6.3	12.3	0.51
Over 2000	5	5.7	9.7	0.59	10	15.6	9.7	1.61
(Over 1000)	(20)	(22.7)	(22.0)	(1.03)	(14)	(21.9)	(22.0)	(1.00)
Total	88	100	100		64	100	100	
USA: South and West only								
200–300	18	26.9	30.5	0.88	7	46.7	30.5	1.53
300–500	18	26.9	26.8	1.00	4	26.7	26.8	1.00
500–1000	14	20.9	20.7	1.01	3	20.0	20.7	0.97
1000–2000	13	19.4	15.9	1.22	0	0	15.9	0
Over 2000	4	6.0	6.1	0.98	1	6.7	6.1	1.10
(Over 1000)	(17)	(25.4)	(22.0)	(1.15)	(1)	(6.7)	(22.0)	(0.30)
Total	67	100	100		15	100	100	

(2) *Specialized service centres* specialize in the provision of a narrower range of intermediate service activities and are subclassified as either '*functional nodal*' (for example, automobile-oriented production services in Detroit) or 'Government and Education' (state capitals, university centres)

(3) *Production centres* are subgrouped into centres specializing in the production of *manufacturing* goods, *industrial–military* output, or *mining–industrial* output

(4) *Consumer-oriented centres* are resort or holiday areas, retirement centres, or are characterized by other primarily consumer-related economic activities.

Noyelle and Stanback then assign the 140 largest US Standard Metropolitan Statistical Areas (SMSAs) to their functional categories based on a cluster analysis of location quotients for the major employment categories. We make use of the Noyelle–Stanback classification and functional assignment to examine the extent to which economic performance is affected by changing urban function in the USA. Data unavailability prevented similar analyses in other countries.

In terms of overall economic performance, consumer-oriented centres were greatly over-represented among good performers (seven good performers, nearly three times the expected number) and under-represented among poor performers (none), while manufacturing centres had the worst performance record (fourteen poor performers), three times the expected number, and no good performers. Functional nodal centres were also very poor performers (twelve performers, three times the expected number, and only three good performers, about 60 per cent of the expected number).

The consumer-oriented areas had the highest median level of total employment growth (1.17 times the national average), while the manufacturing and functional nodal areas had the lowest (0.85 and 0.86 times the national average respectively): see Table 2.6. The industrial–mining, consumer-oriented, and government and education centres all had percentage point unemployment rate increases 30 per cent less than the national average (37, 33, and 30 per cent respectively), followed closely by industrial–military centres (26 per cent less than the national average) and diversified service centres (19 per cent less). Both the manufacturing and the functional nodal centres had median unemployment rate increases substantially above the national average (54 per cent above and 35 per cent above, respectively).

Table 2.6. *Economic Performance and Urban Function in the USA*

Functional type	Median change relative to national average		
	Employment growth	Unemployment rate	No. in sample
Diversified advance service	0.95	0.81	37
Functional nodal	0.86	1.35	22
Government and education	1.01	0.70	19
Manufacturing	0.85	1.54	24
Industrial–military	1.03	0.74	12
Industrial–mining	1.01	0.63	7
Consumer–oriented	1.17	0.67	12

It appears clear that areas specializing in manufacturing production and in providing intermediate services to a single or small number of manufacturing sectors experienced relative economic decline during the 1970s, while consumer-oriented areas in particular have done very well. The diversified advanced service centres, the areas at the centre of the new urban hierarchy, show mixed results: they have had a somewhat slower-than-average growth rate, but an unemployment rate increase less than the national average (though not as much less as several other functional urban groups).

Industrial structure

Does an urban area's past, as reflected in its inherited industrial structure, determine how well its economy performs in the current period? In particular, given the change in national economic structure, did those areas which inherited an economic structure heavily dominated by manufacturing employment perform more poorly than those more heavily dominated by the services?

The results of shift–share analysis (see, for example, Fothergill and Gudgin 1982, 48–68; Hall and Hay 1980, 210–23) suggest that industrial structure by itself accounts for little of the poor performance of urban areas and that the negative local factors consistently and heavily outweigh any negative change in the structural component.

However, the question we ask here is not whether the poor economic performance of urban areas is due to change in the national economic structure—that is, whether an area with a concentration of industries which are growing slowly or declining nationally is inevitably doomed to poor performance—but whether, for whatever reasons, areas heavily dominated by manufacturing at the beginning of the period under study perform worse than other areas. Thus, it may be that areas dominated by manufacturing performed poorly because such areas consistently suffered adverse 'differential shifts', rather than, or in addition to, adverse structural shifts.

For Britain, the USA, and the Netherlands, industrial structure at the beginning of the period did appear to make a difference (see Table 2.7). In each of these three countries the poor-performing urban areas started the period with a much higher percentage of their work-force in the manufacturing sector than did the good performers. In Britain, the good performers began the decade with a median of 28.64 per cent of their workers employed in the manufacturing sector compared to 45.74 per cent for the poor performers. In the USA, the median values were 15.17 per cent for the good performers and 38.70 per cent for the poor performers. This gap continued to exist, although it narrowed, when region was controlled for. In the

Table 2.7. *Performance and Industrial Structure in Britain, the USA, Germany and the Netherlands*

	Median % manufacturing		
	Good performers	Poor performers	Ratio Good : Poor
Britain (1970)	28.64	45.74	1 : 1.60
USA (1970)	15.17	38.70	1 : 2.55
Northeast–Midwest (1970)	28.71	39.40	1 : 1.37
Germany (1978)	48.8	39.4	1.24 : 1
Netherlands (1970)	25.82	36.15	1 : 1.40

Northeast–Midwestern regions of the USA, the group of good performers (in relative terms) started the period with 28.71 per cent of their employees in manufacturing, compared to 39.40 per cent for the group of poor performers. In the Netherlands, the group of good performers had 25.82 per cent of their employment in the manufacturing sector in 1970 compared to 36.15 per cent for the group of poor performers.

In Germany, however, the situation at the national level was reversed, with the good performers actually beginning the period with a higher percentage of manufacturing employment than the poor performers. If we control for region in Germany, however, the expected relationship partially appears. In the two southern regions (Bavaria and Baden-Wurtemburg) the good performers began the period (1978) with a median value of 44.2 per cent of total employment in manufacturing compared to 53.2 per cent for the poor performers. However, in the remainder of the country the good performers started the period with a higher percentage of their total employment in manufacturing (40.2 per cent) than did the poor performers (34.6 per cent).

In both Great Britain and the USA, the relationships between industrial structure and total employment change and between industrial structure and change in unemployment rate were quite strong (see Table 2.8). In Britain, for each additional 1 percentage point of total employment engaged in manufacturing in 1971 there was, on average, a 0.613 percentage point lower growth rate in total employment between 1971 and 1981 ($r = -0.567$), and a 0.111 percentage point higher increase in the unemployment rate ($r = 0.570$).

Table 2.8. *Impact of Percentage of Total Employment engaged in Manufacturing in 1970 on Total Employment Change and Unemployment Rate Change in Britain, the USA, and Germany*

	Total employment change		Unemployment rate change	
	r	b	r	b
Britain	−0.567	−0.613	0.570	0.111
USA	−0.617	−1.478	0.656	0.137
Northeast–Midwest	−0.314	−0.413	0.508	0.118
Germany	0.152	0.069	−0.424	−0.074
Northern region	−0.208	−0.08	−0.183	−0.025
Southern region	−0.341	−0.173	−0.282	−0.495

In the USA, for each additional 1 per cent of total employment engaged in manufacturing in 1970, there was, on average, a 1.478 percentage point lower growth rate in total employment ($r = -0.617$) and a 0.137 percentage point greater increase in the unemployment rate ($r = 0.656$). Thus, in Great Britain, an urban area with 40 per cent of its employment in manufacturing in 1971 could be expected to have 12.3 percentage points less growth in total employment and 2.2 percentage points more of an increase in unemployment rate between 1971 and 1981 than an area which had 20 per cent of its employment in manufacturing in 1971. In the USA, the area with 40 per cent of total employment in manufacturing could be expected to have 29.6 percentage points less growth and 2.74 percentage points more of an increase in the unemployment rate.

In Germany, however, there was essentially no relationship between percentage of manufacturing employment in 1978 and total employment growth between 1978 and 1983. In addition, there was a modest inverse relationship between percentage of manufacturing employment in 1978 and percentage point change in the unemployment rate between 1978 and 1983: urban areas whose industrial structure was heavily dominated by manufacturing had lower increases in unemployment rate than did areas less dominated by manufacturing. This last relationship decreased when region was controlled for, but none the less remained. Thus, it appears that in Germany, unlike in Britain or the USA, an urban industrial structure dominated by the manufacturing sector was not associated with poor economic performance, and, indeed, there is some evidence that it may even have enhanced performance.

While the difference between the German experience and that of the USA and Britain may partly reflect the difference in time periods studied (1978–83 as opposed to the entire decade of the 1970s), it none the less suggests that manufacturing areas are not necessarily doomed to poor economic performance. Even in Britain there were examples of manufacturing areas which performed well. Among the good British performers were several in which manufacturing as a percentage of total employment in 1971 exceeded the national percentage of 36.45 per cent. These included Derby (51.46 per cent), Nottingham (43.87 per cent), High Wycombe (43.13 per cent), and Northampton (40.61 per cent). This phenomenon is explored further in the next section.

Sectoral performance

It has been widely argued that differences in the economic performance of urban areas during the recent past relate primarily to how successful they have been in gaining employment in the rapidly growing service sector. Our data suggest that this is not the case and that, at least during the period under study, the difference between successful and unsuccessful urban economic performance was due at least as much to differences among areas in how well the manufacturing sector performed as to differences in service sector performance.[8] While this appears to be true in the three countries for which data are available (see Table 2.9), it is particularly pronounced in Britain, where there is very little difference (3.9 per cent) in service sector growth between good performers and poor performers, but an enormous

Table 2.9. *Median Percentage Change in Employment by Sector in Britain, the USA, Germany, and the Netherlands*

	Good performers	Poor performers	Difference: Good performers − poor performers
Britain 1971–81			
Manufacturing	−9.1	−32.7	23.6
Services	55.7	51.8	3.9
USA 1970–80			
Manufacturing	58.0	−8.2	66.2
Services	71.6	33.8	37.8
USA: Northeast–Midwest 1970–80			
Manufacturing	−1.8	−14.1	12.3
Services	45.7	32.0	13.7
Germany 1978–83			
Manufacturing	−1.3	−10.1	8.8
Services	15.5	8.9	6.6
Netherlands 1970–80			
Manufacturing	−11	−23	12
Services	n/a	n/a	

difference (23.6 per cent) in employment change in the manufacturing sector. The median value of change in manufacturing employment was a decline of 32.7 per cent for poor performers compared to a decline of only 9.1 per cent for good performers.

Taken as a percentage of 1971 total employment, the British good performers lost 2.27 per cent of their 1971 total employment between 1971 and 1981, through losses in the manufacturing sector, while the poor performers lost 14.25 per cent of 1971 employment through losses in that sector. On the other hand, the good performers gained 10.38 per cent of their 1971 total employment through gains in the service sector, while the poor performers gained 6.85 per cent. Thus, as a consequence of changes in these two sectors, the good performers increased their total employment between 1971 and 1981 by 8.11 per cent, while employment of the poor performers fell by 7.40 per cent—a difference of 15.5 per cent in favour of the good performers. More than three-quarters of that difference (11.98 per cent) was due to performance in the manufacturing sector and little more than one-fifth to service sector performance (3.53 per cent).

Regression analysis of all the urban areas over 200 000 population in Britain supports these conclusions. Each additional 1 percentage point of employment growth in the manufacturing sector was associated, on average, with a change of 0.09 percentage points less in the unemployment rate between 1971 and 1981 ($r = -0.55$). In other words, an area whose manufacturing employment grew by 11 per cent more than another area during

1971–81 would have also had a 1 percentage point lower increase in unemployment rate over that period than the slower-growing area. On the other hand, each additional 1 percentage point in service sector employment growth was associated, on average, with only a 0.02 percentage point lower change in the unemployment rate, and the correlation was much lower ($r = -0.12$).

In addition, the rate of change in manufacturing employment in British urban areas was very highly related to the rate of change in total employment. Each additional 1 percentage point increase in manufacturing employment was associated, on average, with an additional 0.69 percentage point increase in total employment ($r = 0.80$). The increase in total employment for each additional 1 percentage point of service employment was lower (0.50), and the correlation was much lower ($r = 0.49$).

In Germany, each additional 1 percentage point of manufacturing employment was associated with a 0.14 percentage point lower increase in unemployment ($r = -0.43$), while each additional percentage point of service sector employment was associated with only a 0.07 percentage point lower increase in the unemployment rate ($r = -0.27$). Each additional 1 percentage point of manufacturing employment was associated with an additional 0.755 percentage point increase in total employment ($r = 0.88$), while each additional percentage point of service sector employment was associated with an increase of only 0.49 percentage points in total employment ($r = 0.68$).

However, the results for the USA do not present so clear-cut a case. For the country as a whole, the contrast between the good performers and poor performers followed the same pattern as in Britain, with growth rate differences between the two groups much greater in manufacturing employment than in services. However, when controlled for region, the growth rate differences between the two sectors disappear. In addition, when taken as a percentage of 1970 employment, approximately 60 per cent of the greater growth by the good performers was due to differences in the service sector compared to 40 per cent due to differences in the manufacturing sector (both for the country as a whole and for the Northeast-Midwest rgion).

Regression analysis for US urban areas indicates that change in manufacturing employment and change in service sector employment had nearly equal impacts both in terms of their effect on change in unemployment rate and in terms of their effect on total employment growth between 1970 and 1980. Each additional 1 percentage point of employment growth in manufacturing was associated, on average, with 0.035 percentage points less of an increase in the unemployment rate ($r = -0.540$), exactly the same impact as a 1 percentage point increase in service sector employment ($r = -0.44$). An additional 1 percentage point of manufacturing employment was also associated with, on average, an additional 0.66 percentage points of total employment growth ($r = 0.87$), while an additional 1 percentage point of service sector employment growth was associated with an additional 1.06 percentage points of total employment growth ($r = 0.91$).

These results seem, at a minimum, to cast strong doubt on the assertion that the service sector was the key to successful urban economic performance during the 1970s. In Britain, it clearly was the performance of the manufacturing sector which distinguished good from poor urban economic performers, and in the USA, the manufacturing sector and the service sector were roughly of equal importance.

Indeed, these results are quite surprising. What could account for the more robust impact of changes in manufacturing employment compared to service employment on unemployment rates and total employment growth in Britain? First, it is possible that manufacturing jobs have more powerful local linkage effects than do service jobs. If this were so, an additional manufacturing job would result in a greater increase in employment among *local* firms supplying its inputs than would an additional service sector job. Secondly, if manufacturing jobs paid higher wages than service sector jobs, the local multiplier effect of additional manufacturing jobs on local income and employment could be expected to be higher than that of an additional service sector job.

It is also possible that the effect of changes in the number of service sector jobs on local unemployment rates is mitigated through changes in net migration and activity rates to a greater extent than is the case with changes in the number of manufacturing jobs. This is consistent with Gordon's finding (Gordon 1984, 50) with respect to the London area that 'The tendency for the effects of differential employment change to be dispersed through surrounding areas has been strongest in the case of service (or non-manual) employment', presumably because of the greater mobility of service sector workers.

None of the above explanations, however, would help to account for the difference in the relative impact of changes in manufacturing and service sector jobs in Britain as compared to the USA unless it is the case that local linkages, manufacturing–service sector pay differentials, or the relative mobility of manufacturing and service sector workers also differ among urban areas in the two countries. A more interesting possibility is that the impact of sectoral employment change on economic performance is not symmetrical with respect to gains and losses, and that the impact of a change in manufacturing employment compared to an equivalent change in service employment is greater when manufacturing employment is being lost (as is the case with most British cities) than when it is being gained (as is the case with many US cities). Overall manufacturing employment declined by 24 per cent in Britain during the 1970s, compared to an increase of 5 per cent in the USA.

This substantial decline of manufacturing employment in Great Britain compared with the USA, combined with the much slower growth of service employment (22 per cent compared to 41.4 per cent) also helps to explain why change in manufacturing employment rather than service employment had a greater impact on differentiating between good and poor urban economic performance in Britain than in the USA. The scope for differential performance among local areas undoubtedly increases with the extent of national change. Thus, when service sector employment increases by 41 per cent nationally, as in the USA, it is more likely that the variation among local areas in service sector employment change will be greater than when it changes by only 22 per cent, as in Britain. In short, the better national performance in the service sector permitted expansion of service sector jobs a greater role in differentiating between good and poor performers locally. Similarly, there is more scope for variation among urban areas in manufacturing employment with a 24 per cent national decline, as in Britain, than with a 5 per cent national change as in the USA.

The results permit us now to place in better perspective the findings of the previous section. It is not how *dominant* was the manufacturing sector at the beginning of the period (that is, industrial structure) which determined urban economic performance, but how well that sector *performed* over the course of the period. Thus, urban area economies with a high percentage of employment in manufacturing performed well if their manufacturing sector performed relatively well (for example, Northampton, High Wycombe, and Derby in Britain).

However, was there a tendency for the manufacturing sector to perform poorly in areas whose industrial structure was heavily dominated by manufacturing, thus resulting in a poor overall economic performance? In the USA and Great Britain, such a tendency did appear to exist: the performance of the manufacturing sector, as measured by the rate of growth in manufacturing employment during the 1970s, was related to the dominance of the sector, as measured by manufacturing employment as a percentage of total employment ($r = -0.57$ in Great Britain and -0.58 in the USA). Those urban areas with a high proportion of their employment in the manufacturing sector at the beginning of the period tended to experience a *higher* rate of decline in manufacturing employment and thus a poor overall economic performance. However, when region was controlled for in the USA the correlation dropped sharply ($r = -0.13$ for the Northeast–Midwest region).

In Germany, on the other hand, there was very little correlation between manufacturing dominance and manufacturing performance ($r = 0.20$). Unlike in Britain or the USA, areas in Germany whose industrial structure was heavily dominated by manufacturing at the beginning of the period were as likely as other areas to experience above-average growth in manufacturing employment over the period studied.

The data presented in this section emphasize the importance of the performance of the manufacturing sector in determining overall economic performance and particularly in accounting for differences between the best and worst group of performers. The data unfortunately do not tell us why or how some areas performed better in manufacturing than other areas. For this task, a more detailed analysis is necessary. To what extent did urban area manufacturing performance simply reflect industrial structure (that is, areas specializing, at the beginning of the period, in those manufacturing subsectors which performed well nationally over the period)?[9] If, for example, areas with a lower proportion of employees in the manufacturing sector also had a tendency to specialize in subsectors which were growing faster nationally than those characterizing areas more heavily dominated by manufacturing employment, it would explain our findings that the manufacturing sector in the former areas outperformed that in the latter ones. Shift–share analysis disaggregated at least to two- or perhaps even three-digit Standard Industrial Classification (SIC) codes would be required for this type of analysis.

If, however, previous research on shift–share analysis can be believed, and differences in manufacturing performance among urban areas do not simply reflect proportionate changes in the shares of national industrial structure, then the variations in performance are due to variations in the differential shifts among areas. What accounts for these variations in differential shifts, and why would they be more adverse in areas heavily dominated by manufacturing? Did the successful areas have a capital stock of more recent vintage within any given manufacturing subsector? Did the successful areas have characteristics which permitted more profitable production—greater labour productivity, lower factor costs—than their less successful counterparts? Did the less successful areas suffer from constraints—such as the lack of available land—that the more successful areas did not experience? Did the successful areas benefit from the location of major universities and research institutes or of highly skilled labour? What role did public policy play?

Nor do the data provide information on the *process* by which urban areas experienced change in the manufacturing sector. Did the relatively more successful areas succeed by retaining existing manufacturing firms more successfully than did their less successful counterparts or were they more successful because they were better able to generate new firms? If the latter, to what extent were these new firms in the same manufacturing subsectors as previously and to what extent did the area's mix of manufacturing activities change?

Skill levels

The decline in lower-skill manufacturing jobs and the rise in importance of employment associated with more complex manufacturing processes and with knowledge-intensive services suggest that labour-force skill levels may be an increasingly important determinant of urban economic performance. Do urban areas with more highly skilled labour forces perform better than areas with lesser skilled labour forces?

To examine this question we first compared the group of good overall economic performers to the group of poor economic performers in terms of the occupational skills of the labour force. In both the USA and Great Britain, the good performers had a higher percentage of their labour force in professional occupations and a lower percentage in unskilled and semi-skilled occupations. In the USA, the pattern in the Northeast–Midwest region was similar to that of the nation as a whole (see Table 2.10). In Britain, for example, the good performers had a median of 36.6 per cent of their labour force classified as professionals compared to 25.5 per cent for poor performers, but only 18.8 per cent of their labour force was semi-skilled or unskilled, while nearly a quarter of the labour force of the poor performers was in these categories. This is consistent with the findings of Begg *et al.* (1986) who find a correlation of approximately −0.49 between low skilled groups (socio-economic groups (SEGs) 7, 10, and 15) in 1981 and employment change between 1971 and 1981, and a correlation of 0.3 between higher skilled groups (SEGs 1, 4, and 13) and employment change.

Table 2.10. *Economic Performance by Occupation in Britain and the USA*

	Median % of labour force	
	Professional	Semi-skilled and unskilled
Britain		
Good performers	36.6	18.8
Poor performers	25.5	24.9
Difference	11.1	−6.1
USA		
Good performers	28.0	13.3
Poor performers	23.9	22.9
Difference	4.1	−9.6
USA: Northeast–Midwest		
Good performers	27.5	17.2
Poor performers	22.8	24.0
Difference	4.7	−6.8

Note: Professionals are defined in Britain as social classes I and II, and in the USA as the following occupational categories: managerial, professional, technical, and related support occupations. *Semi-skilled and unskilled workers* are defined in Britain as social classes IV and V, and in the USA as the following occupational categories: operators, fabricators, and labourers. Definitions are not comparable across countries.

In addition to occupational characteristics, we also examined education levels as an indicator of labour-force skills. While data are not available for Britain, data for the USA indicate that the good-performing urban areas have better-educated labour forces than the poor-performing areas and that these differences are not due to regional variations in education levels. Thus, in the good-performing areas, 19.6 per cent of the population above the age of twenty-five had four or more years of college education compared to only 13.6 per cent of those in poor-performing areas (the comparable figures for the Northeast–Midwest region were 19.4 per cent and 13.1 per cent).

Factor costs

It has frequently been argued that economic performance is related primarily
to differences in factor costs, and especially in labour costs. Areas with relatively high labour costs, it is contended, will not attract investment, owing to
lower returns on capital, and thus will suffer from poor economic performance. In Britain, Tyler *et al.* (1984, 47–56) found a profit gradient for
manufacturing firms in a range of industries in southern England, with
profits increasing with distance from London. Most of the difference was
attributable to wage and salary costs. However, Fothergill *et al.* (1985,
10–11), using a different methodology and data, found that there were relatively small differences in costs across areas and that these differences were
unlikely to be sufficient to affect relative economic performance. They argue
that, given the prevalence of national wage settlements, local wage costs bear
little or no relation to local labour-market conditions.

In the USA, however, evidence exists to suggest that the much greater
spatial variation in wage costs does make a difference in terms of economic
performance. In an econometric study of new births of manufacturing firms,
for example, Carleton (1979, 15) concludes that 'Wages matter a great deal
in explaining births of single establishment firms. For every one percent
decrease in wages, new births increase by one percent. For branch plants, the
wage effect, though probably large, could not be estimated with much
precision.'

We compared relative wage rates in three occupational categories for the
two groups of US urban areas (good and poor performers) in 1976 and
again in 1982. In terms of unskilled plant jobs, the good-performing areas
had lower wage rates than did the poor-performing areas in 1976, and by
1982 the differences in wage rates between the two groups had increased.
This pattern held true for the Northeast–Midwest region as well as for the
entire nation. Thus, in the Northeast–Midwest region—the region traditionally with the highest wages in the country—the good-performing
areas had unskilled plant wage rates 1.3 per cent above the national average in 1976, while the poor performers had unskilled plant wage rates 9.5
per cent above the national average. By 1982 the gap had increased: the
comparable figures were 1.4 per cent and 11.4 per cent above the national
average.

By contrast, there was little difference in wage rates for skilled maintenance jobs between the two types of areas in 1976 (1.1 per cent above the
national average for poor performers in the Northeast–Midwest, compared
with 0.2 per cent above the national average for good performers), and the
slight difference had actually narrowed by 1982. This convergence effect was
even more pronounced in terms of wages for office clerical jobs. In these jobs,
wage rates were 2.6 percentage points above the national average in poor-
performing areas in 1976 compared to 1.8 percentage points below the

national average in good-performing areas. By 1982, however, the differential had completely disappeared.

These results are as expected for skilled maintenance and office clerical jobs. The poor performers over the course of the decade had higher wage rates than did the good performers near the decade's mid-point, but the poor economic performance triggered an adjustment process which, by the early 1980s, had resulted in relative wage declines in these areas. However, a similar adjustment process did not seem to occur with respect to unskilled plant wage rates; indeed, the differences between good and poor performers actually increased between 1976 and 1982.

In addition to wages, the cost and availability of other factors may contribute to differences in urban economic performance. Thus, Fothergill *et al.* (1985) argue that the relative availability of usable land is the primary factor accounting for differences in economic performance between large and small urban areas. We were, however, unable to collect data to enable us to examine this hypothesis.

Amenities

The oft-cited decline in the importance of transportation considerations in the location decision has made firms more 'footloose'. As a consequence, considerations relating to the pleasantness of the working and living environment may now play a more important role, particularly for the growing number of firms dependent upon recruiting highly skilled professional and technical labour. If such labour chooses to locate in amenity-rich locations, their presence may attract firms to these locations. Thus, for example, Birch (1984, 13) contends that his research indicates 'Employers are seeking out better educated and trained labour forces (not cheaper ones), higher service delivery levels (not lower taxes) [and] higher quality of life (not lower land and building costs) . . . '

Amenity considerations which have been cited in this regard include climate, physical beauty, recreational opportunities, air and water quality, athletic and cultural institutions and events, quality of public services (particularly education), and the level of crime. It is likely that some mix of all of these, plus other characteristics, contributes to perceptions of an area's quality of life. As a consequence, obtaining an objective measure of an area's composite quality of life appears impossible (indeed the most appropriate measure would probably be subjective in the form of an opinion survey). Instead we have selected one measure—the crime rate—for examination, both because it is frequently cited as having a major impact on people's perception of quality of life and because it is one of the few measures for which data are available at the metropolitan area level, albeit only for the USA.

Our data indicate that at the national level the good-performing areas in the USA had a substantially *higher* crime rate than did the poor-performing areas (7230 crimes per 100 000 residents compared to 5575). The difference completely disappears, however, when region is controlled for: within the Northeast–Midwest region there is essentially no difference between the crime rates of good- and poor-performing urban areas.

The Performance of National Urban Systems

Our study to this point has been concerned with variation in the performance of urban economies *within* countries. If we wish to compare the performance of urban areas *across* countries we must first control for the performance of the national economy, since the economic performance of urban areas in a country obviously reflects, to some significant extent, the performance of the national economy as a whole. In order to control for national economic performance we express the two performance variables for each urban area economy—change in unemployment rate and change in total employment—relative to the national economic performance. Thus, if the local unemployment rate increased by 5 percentage points and the national rate had increased by 4 percentage points, the change in the local unemployment rate would be expressed as 1.25.

Using this methodology, we can assess the performance of the urban systems (defined as all urban areas with 1970 populations in excess of 200 000) in each country and then make comparisons across countries. We find that in all four countries (the USA, Britain, Germany, and the Netherlands), the urban systems suffered higher unemployment rate increases than did the nation as a whole. Comparing urban systems across countries, we find that the urban systems of the Netherlands and Britain had the worst performances of the four countries, while Germany's had the best. However, in terms of total employment growth, the urban system outperformed the national economy in all of the four countries except the USA. This was particularly true of the Netherlands, where the total employment index for urban areas in 1980 was 1.18 times that of the national total employment index, but was true as well of Britain (1.03) and Germany (1.01) (see Table 2.11).

How do the countries compare with respect to variations in performance *within* their urban systems? Was there a greater variation in economic performance among urban areas in Britain than the other countries? To answer this question we must look at the standard deviation for each country—that is, the extent to which the performance of the urban areas in each country varied around the mean performance for that country.

In absolute terms there was a greater variation around the mean value for unemployment rate change in Britain than in the other countries: the standard deviation for all British urban areas was a 2.20 percentage point change compared to 1.83 in the USA and 1.56 in Germany. The standard deviation

Table 2.11. *Urban Economic Performance of National Urban Systems in Britain, the USA, Germany, and the Netherlands*

Country	No. of urban systems	Unemployment rate change index[a]	Total employment change index[b]
USA	151	1.04	0.96
Britain	61	1.05	1.03
Germany	61	1.02	1.01
Netherlands	12	1.06	1.18

[a]Unemployment rate change index calculated as median percentage point change in urban area unemployment rates divided by percentage point change in national unemployment rate.
[b] Total employment change index calculated as urban area total employment index in final year (initial year = 100) divided by national total employment index in final year (initial year = 100).

for employment growth was larger in the USA (27.05 percentage points) than in Britain (12.23) or Germany (4.10).

However, to obtain a more meaningful comparison, it is necessary to examine the extent of variation around the mean value, after first controlling for differences in mean values among the nations. To do this we calculated the coefficient of variation (the standard deviation divided by the mean for each country—CV) and found that there was much greater variation among US urban areas with respect to change in unemployment rate (CV = 0.70) than in either Britain (CV = 0.28) or Germany (CV = 0.27). Similarly, there was greater variation among US urban areas with respect to employment growth (CV = 0.20) than among British urban areas (CV = 0.12) or German urban areas (CV = 0.04). In short, the extent of variation in performance among urban areas was much greater in the USA than in the other countries, with British urban areas experiencing about the same variation in unemployment rate change as German areas, but a greater variation in growth rates (see Table 2.12).

Table 2.12. *Extent of Variation in Urban Economic Performance by Country for Britain, the USA, and Germany*

Country	Change in unemployment rate			Total employment growth index[a]		
	Mean[b]	Standard deviation	Coefficient of variation	Mean[b]	Standard deviation	Coefficient of variation
USA	2.62	1.83	0.70	135.80	27.05	0.20
Britain	7.94	2.20	0.28	99.77	12.23	0.12
Germany	5.73	1.56	0.27	100.96	4.10	0.04

[a] 1971 = 100.
[b] For all urban areas with population in excess of 200 000.

Summary

(1) The study utilized two measures of urban economic performance: change in the unemployment rate and growth in total employment. In all three of the countries with well-developed urban systems (Britain, the USA and Germany), relatively high employment growth was associated with a relatively low increase in the unemployment rate.

(2) In three of the four countries (Britain, Germany, and the Netherlands), those urban areas with the highest unemployment rates at the beginning of the period had the poorest performance over the course of the period. At the end of the period they were relatively even worse off than when they had started. This was not true in the USA where there was little relationship between initial standing and performance over the period.

(3) There was a very strong regional dimension to urban economic performance. In Britain, good performers were highly concentrated in the South and East and poor performers in the North; in the USA, good performers were concentrated in the West and poor performers in the Northeast and Midwest; and in Germany good performers were heavily concentrated in the southernmost two states, while poor performers were concentrated in the northernmost ones.

(4) In all countries the very largest urban areas appeared to be disproportionately poor performers. However, for size categories below 1 million, there did not appear to be a gradient effect: smaller urban areas were no more likely than larger ones to be good or poor performers (Germany was a mild exception to this generalization and a slight size gradient did appear to exist for Britain with respect to rate of total employment growth, but not with respect to change in the unemployment rate).

(5) In the USA—the only country for which analysis was possible—urban areas specializing in manufacturing production and in providing intermediate services to a single or small number of manufacturing subsectors experienced the poorest economic performance during the course of the 1970s, while consumer-oriented areas performed particularly well. Areas specializing in government and education, in industrial–defence employment, and in diversified advanced services also performed substantially better than the nation as a whole in terms of changes in the unemployment rate, but only marginally better—or, in the case of diversified advanced services, marginally worse—with respect to the rate of total employment growth.

(6) In three of the four countries (Great Britain, the USA, and the Netherlands) industrial structure of urban areas at the beginning of the period appeared to be closely related to the performance of urban areas over the period. Urban areas experiencing poor economic performances started the period with a much higher percentage of their work-force in the manufacturing sector and a much lower proportion in the service sector than did the

good performers. However, in Germany the situation was reversed: the good performers started the period with a higher proportion of their work-force in manufacturing than did the poor performers (though when region was controlled for there was little difference in industrial structure between good and poor performers).

(7) Differences in performance between good and poor performers were due in all countries at least as much to how well the manufacturing sector performed as to how well the service sector performed. In Britain, the differences were due overwhelmingly to the performance of the manufacturing sector, which accounted for more than three-quarters of the difference between good and poor performers in employment change in the combined manufacturing and service sectors. Indeed, it was the performance of the manufacturing sector rather than its dominance which was most convincingly related to economic performance. Urban area economies with a high percentage of their employment in manufacturing performed well if their manufacturing sectors performed well. However, in the USA and Britain, there was a tendency for manufacturing employment to grow at a slower pace in areas whose industrial structure was heavily dominated by manufacturing and at a more rapid rate in those areas with a relatively low proportion of manufacturing employment. In Germany, though, there was little correlation between manufacturing dominance and manufacturing performance.

(8) Urban area performance appeared to be related to labour-force skill level in both the USA and Britain, and to education levels in the USA (data for Britain were unavailable).

Conclusions

There are a variety of implications which might be drawn from these findings, particularly when placed within the context of other research:

(1) The regional context of urban economic performance in all countries was quite striking.[10] This suggests that there is a strong regional component to urban economic performance and that efforts to improve the performance of urban economies must go hand in hand with measures to improve the performance of the regional economy of which the urban area is a part.

(2) The finding that differences among areas in the performance of their manufacturing sector had a greater impact on differentiating between good and poor economic performers than differences in service sector performance suggests that the gulf between good and poor performers may lessen as the run-down in manufacturing employment finally comes to a halt. Thus, if the problem is seen as the *extent of variation* in the performance of Britain's urban economies and the size of the gulf between good and poor performers, there is good reason to believe that, even

without effective policy intervention, the variation and gulf are likely to diminish over time.

(3) The finding that changes in manufacturing employment had a more robust impact on an area's unemployment rate and employment growth than changes in service employment should be treated with great caution in policy terms. In particular, it does not necessarily suggest that urban areas should redouble their efforts to attract traditional manufacturing activity. Even though the pay-off to such efforts might be quite high if they were to succeed, the ability to increase employment in this kind of activity through local efforts may be quite limited, while the ability to increase service sector employment through local efforts (including public sector employment) may be much greater. On the other hand, the findings also suggest that broad and sweeping conclusions about deindustrialization and the dominance of the service sector are premature, that areas with a high proportion of their employment in manufacturing are not necessarily doomed to poor economic performance, and that those areas which are able to perform well in the manufacturing sector (most likely because they are competitive in more complex 'high technology' sectors) are likely to experience very favourable overall economic performance.

(4) The finding that good urban economic performance was related to labour-force skill and education levels suggests the need for greater emphasis on improved access to higher education and vocational training for the general work-force—an area in which Britain lags badly behind the USA and Germany.

(5) The findings that among the most successful economic performers in the USA were areas specializing in employment related to government, education, and industrial–defence suggest that government intervention with respect to the placement of public facilities and the spatial targeting of procurement can have an important impact on the performance of local economies.

The research results, taken together, indicate that there are indeed strong forces at work affecting urban economies across the advanced Western nations. They also suggest that there are some important differences in the nature of these forces or in the effects they have from country to country. However, as the above discussion indicates, governments are not completely devoid of means for affecting urban economic performance, should they choose to make use of them.

Notes

1. Cheshire *et al.* (1986) point out that 'The fully systematic international studies (Hall and Hay and van den Berg *et al.*) only analyze employment data briefly.' Hall and Hay (1980, Ch. 5) do attempt to explain total metropolitan area

employment change across European metropolitan areas from 1960 to 1970, through a restricted set of independent variables: population level, total employment level, employment density, employment concentration in the core, latitude, and city edge defined as 1870 population as a percentage of 1970 population. Using simple correlation analysis, they found no strong consistent relationship between any of these variables and total employment change. Hall and Hay (1980, Ch. 6) also perform a shift–share analysis for selected metropolitan areas for the 1960–70 period. They found that in most cases the differential (local) shift was more important than the structural.

2. Since Northern Ireland was excluded, we refer to Great Britain rather than the United Kingdom.
3. For precise specifications, see Appendix 2.
4. It must be emphasized that this study was limited to the 1971–81 period. Had it covered the 1951–81 period, Glasgow would have surely fallen into the class of poor performers. Its 'mixed' performance between 1971 and 1981 simply reflected, to a large degree, the extent of its economic decline between 1951 and 1971.
5. A rigorous analysis would have involved efforts to control for interaction among the various possible explanatory variables (multicollinearity) through use of multiple regression analysis or through cross-tabulation in which the relationship of one variable to another could be observed for each category of a third (or additional) control variable. Multiple regression analysis of this sort was beyond the scope of this project, both for technical and for resource limitation reasons. Controls through cross-tabulation were, in most cases, limited by the relatively small number of urban areas in each urban system and thus the very small number of areas in some cells. We did impose some controls through cross-tabulation for region (by dividing each country into two regions), but most of the analysis does not attempt to impose controls and the regression analysis is simple regression rather than multiple regression.
6. See also Keeble *et al.* (1983).
7. These findings are consistent with those of Greenwood (1981, 62) who, using rank-order correlation, found no relationship between SMSA size and employment growth between 1950 and 1960 and between 1960 and 1970. Greenwood also found that intermediate-sized metropolitan areas experienced higher rates of employment growth than did areas in higher and lower size categories.
8. See Appendix 3 for definition of service sector in each country.
9. Bradbury *et al.* (1982, 91) found in a regression analysis of US SMSAs that SMSAs with a favourable industrial mix (a high concentration of industries with fast national growth rates) grew faster than did other areas.
10. This corresponds with the findings of Buck and Gordon (Chapter 4 of this volume) that local employment growth had a favourable impact on local unemployment rates in growing regions of England (the South), but not in declining regions (the North).

References

Begg, I., Moore, B., and Rhodes, J. (1986), 'Economic and Social Change in Urban Britain and the Inner Cities', in V. A. Hausner (ed.), *Critical Issues in Urban Economic Development*, Vol. 1, Oxford: Clarendon Press.

van den Berg, L., Drewett, R., and Klaasen, L. (1982), *Urban Europe: A Study in Growth and Decline*, Pergamon Press.

Birch, D. (1984), 'The Changing Rules of the Game: Finding a Niche in the Thoughtware Economy', *Economic Development Commentary*, 8(4), 12–16.

Bradbury, K. L., Downs, A., and Small, K. A. (1982), *Urban Decline and the Future of American Cities*, Washington, DC: The Brookings Institute.

Carleton, D. W. (1979), 'Why New Firms Locate Where They Do: An Econometric Model', in W. C. Wheaton (ed.) *Interregional Movements and Regional Growth*, Washington, DC: The Urban Institute Press, 13–50.

Cheshire, P., Hay, D., and Carbonaro, G. (1986), *A Review of Literature on Urban Problems in Europe*, Luxemburg: Office for the Official Publications of the European Communities.

Coombes, M. G., Dixon, J. S., Goddard, J. B., Openshaw, S., and Taylor, P. J. (1981), 'Approximate Areas for Census Analysis: An Outline of Functional Regions', *Discussion Paper 41*, Centre for Urban and Regional Development Studies, University of Newcastle upon Tyne.

Fothergill, S. and Gudgin, G. (1982), *Unequal Growth: Urban and Regional Employment Change in the UK*, London: Heinemann Educational Books.

Fothergill, S., Kitson, M., and Monk, S. (1985), *Urban Industrial Change*, Department of the Environment Inner Cities Research Programme, No. 11.

Gordon, I., (1984), 'Unemployment in London', *London Project Working Paper No. 5*, Urban and Regional Studies Unit, University of Kent.

Greenwood, M. J. (1981), *Migration and Economic Growth in the United States*, New York: Academic Press.

Hall, P. and Hay, D. (1980), *Growth Centres in the European Urban System*, London: Heinemann Educational Books.

Keeble, D., Owens, P. L., and Thompson, C. (1983), 'The Urban–Rural Manufacturing Shift in the European Community', *Urban Studies*, 20, 405–18.

Noyelle, T. and Stanback, T. M. (1984), *The Economic Transformation of American Cities*, Totowa: Rowman and Allenheld.

Tyler, P., Moore, B., and Rhodes, J. (1984), 'Geographical Variations in Industrial Costs', *Discussion Paper No. 12*, Department of Land Economy, Cambridge University.

Appendix 1: Urban Area Definitions

The USA: Standard Metropolitan Statistical Areas (SMSAs) based on the 1980 Census definitions.

Britain: Functional Regions as defined by the Centre for Urban and Regional Development Studies, University of Newcastle upon Tyne (Coombes *et al.* 1981). These include both sub-dominant functional regions of the dominant metropolitan functional regions and urban free-standing functional regions. These regions represent 'urban-centred labour market areas', based on urban cores and surrounding areas based on commuting criteria.

Germany: Functional Urban Regions (FURs) as defined by the Regional Planning Ministry. These regions are defined in terms of commuting patterns to constitute the large labour market surrounding a major urban centre.

The Netherlands: 'COROP' regions defined for regional policy purposes. There are forty COROP regions, covering the entire country. The highly urbanized COROP regions are more extensive than the SMSA or FUR concept; however, it is questionable how much sense urban areas defined by travel-to-work patterns make in a small highly urbanized country such as the Netherlands.

Appendix 2: Definitions of Performance Groups

For the Netherlands, there were not enough cases to classify according to the procedure in Table 2.A1. Instead, the twelve urban areas were rank-ordered on each of the two performance indicators (unemployment rate change and total employment change) and an average rank-order score derived. The top five average rank-ordered areas were designated as good performers, the bottom five as poor performers, and the two areas which tied for sixth place (Amsterdam and Arnhem) as mixed performers.

Table 2.A1. *Definitions of Performance Groups for Britain, the USA, and Germany*

Country	Good performers		Poor performers	
	Change relative to national		Change relative to national	
	unemployment rate	total employment	unemployment rate	total employment
USA	<0.81	>1.04	>1.25	<0.91
Britain	<0.96	>1.00	>1.06	<1.00
Germany	<0.95	>1.02	>1.05	<1.00

Appendix 3: Definition of Variables

Inevitably in cross-national research the definition of important variables frequently is not consistent across countries, rendering direct comparisons difficult. We have attempted to reduce the problems by avoiding direct cross-country comparisons of variables at a single point in time (for example, comparing the 1980 unemployment rate in the USA to that of Britain) and instead have compared changes over time in both countries (for example, comparing changes in the unemployment rate between 1970 and 1980 in the USA to those in Britain), thus assuming that definitional differences will have less of a distorting impact on rate of change than on level. Nevertheless, there are important definitional differences the reader should be aware of, the most important of which are presented below.

Unemployment rate

In the USA, the unemployment rate is defined as the number of those who live in an area and are unemployed and seeking work as a percentage of the total labour force (employed plus unemployed) living in the area. The definition is thus totally residentially based: the location of work is irrelevant. Data are gathered through sample surveys. In the European countries, however, the unemployment rate is defined as the number of those living in an area who are unemployed and seeking work as a percentage of the unemployed plus those working in the area. In short, the denominator includes non-residential commuters into the area and excludes residents who commute outside of the area to work. Data on unemployed are gathered from registration at local unemployment offices.

The differences in definition make unemployment rate comparisons across countries difficult where commuting to or from the area is an important component of employment (that is, in relatively tightly defined areas such as cities or neighbourhoods). However, in FURs, which are defined in terms of commutation patterns, the importance of commuting into or out of the FURs should be relatively small, and unemployment rates should be measuring approximately the same phenomenon in the four countries.

The service sector

Data are provided in broad categories which are not strictly comparable from country to country. We have selected categories which focus on, to the extent possible, traded or exportable services rather than local services (retail and wholesale trade are not included in the service sector in any country). In the USA, the SIC categories included in the service sector are: finance, insurance, and real estate; services (business services, professional and related services, personal services, entertainment, and recreation); and government. In Britain, it includes: finance, insurance, and banking; professional, scientific, and other services; and public administration and defence. In Germany, the categories included in the service sector are credit and insurance, services, non-profit organizations, and government and social services.

Occupational skill

In each country an effort was made to distinguish the least skilled and the most skilled employees. Again, these categories are not strictly comparable across counties. In the USA, the most skilled employees are in the classification of managerial, professional, technical, and related support workers, while the least skilled are in the operators, fabricators, and labourers classification. In Britain, the most skilled categories are professional and intermediate workers (social classes I and II), while the least skilled are semi-skilled and unskilled workers (social class categories IV and V).

3

The Changing Economic Role of Britain's Cities

Iain Begg and Barry Moore

Introduction

Britain's cities are undergoing a major structural transformation. This transformation is the outcome of the complex interplay of social, economic, technological, and institutional changes which combine to influence the location of population and economic activity across the urban system. In the postwar period, improvements in communication, the growing importance of service sector activity in the economy, technological progress in the production of manufactured goods, declining rates of population increase, and changing residential preferences are some of the factors contributing to the now-familiar stagnation and decline of many of Britain's major conurbations and large cities, and to the growth of the smaller towns and urban hinterland. These trends pose critical questions about the role of many of the great cities established at the time of the Industrial Revolution and whose prosperity was based on industries such as shipbuilding, coal, iron and steel, and textiles. Indeed, some cities have fared so badly in recent years that their survival as economic entities is called into question. At the same time the smaller towns and cities, particularly in the southern half of the country, are facing often-difficult problems of adjustment to a more rapid growth of population and economic activity.

The objective of this chapter is to document the performance and changing economic role of the 100 largest urban areas in Britain and to begin to explore the reasons for the observed variations in city performance. Such an analysis will provide a valuable context within which to place the much more detailed studies of the Inner City Research Programme.

The work reported here is based entirely on employment data, not because this is necessarily the best measure of activity, but because it provides the greatest scope for the required disaggregation. The source of the data is Department of Employment records of employment by employment exchange area. These show males and females separately and give a comprehensive breakdown by Minimum List Heading (MLH).

The results presented in this chapter are based on data from the 1971 and 1981 Censuses of Employment. With around 100 MLHs for each of about 800 exchange areas, the basic data are very voluminous, and appropriate computer programs have been written, as part of this project, to organize and manipulate the data. Moreover, because the boundaries of exchange areas in 1971 are not always identical to those in 1981, these have had to be carefully matched in order to obtain consistent boundaries. For a small number of cities, it proved to be impossible to match boundaries completely, but they have nevertheless been retained in the analysis, though not used in change variables.

As explained in more detail below, 'cities' have been defined in terms of entire employment exchange areas, as there is no firm basis on which to split up an exchange area. This means in some cases that there is some rural hinterland within the definition of a city. But, as the data refer to jobs rather than to residents, it is considered unlikely that this brings in much bias to the results.

Industry groupings have been constructed by allocating MLHs to nine groups which are designed to reflect industry and labour demand characteristics as well as going some way towards capturing location factors. A brief description of this is given below.

Selection of cities

The starting point for the selection of cities was the 1981 Census publication *Key Statistics for Urban Areas* (Office of Population Censuses and Surveys 1985). This volume reports Census results for all urban areas, with continuously built-up areas regarded as urban units. In most cases, these have been adopted in the present study, although in a few instances we have changed these definitions. Thus, in the Census definition, Liverpool and Birkenhead are classified separately, but they have been merged in this analysis. A similar policy was adopted in relation to the Glasgow area; however, in some other metropolitan areas 'fringe' towns have been separated.

This was especially the case around London, where the ring of cities used in this chapter is somewhat arbitrary. Essentially, what has been done around London is the creation of pseudo-towns on each of the main road arteries out of the capital. Thus, we have Harlow (M11), Hatfield (A1), Crawley (A21), etc.

A further factor in delineating city boundaries was the need to match Census-based definitions with employment exchange area borders. For larger cities, this was a minor problem, but some smaller areas needed a bit of juggling.

One other point about boundaries is that some 'cities' are in fact amalgamations of separate, smaller settlements. Dearne Valley, for example, is not an independent unit but consists, instead, of a string of small towns and pit villages. Nevertheless, they are all sufficiently close together to have some degree of coherence as a labour market and a community.

Altogether, then, some definitions may be slightly questionable, but, for the vast majority, the actual and employment exchange area boundaries are tolerably accurate. The intention, approximately, was to disaggregate to settlements of over 80 000 inhabitants, and this yielded a list of 104 cities amounting to over three-quarters of the population and employment in Great Britain. As centres of employment, some cities do lie below the target level, largely because the employment exchange area boundaries are narrower than the actual settlements. This, however, poses only minor problems.

Industrial aggregations

There have now been several departures from the traditional industrial clas-sifications of primary, secondary, and tertiary industries developed by Fisher and Clark in the 1930s, although to a significant extent this classification still dominates much conventional applied economic analysis. In large measure, the reclassifications reflect the changing composition of national output, notably the ongoing shift of mature industrial economies towards services and away from manufacturing. Gershuny (1983) and Gershuny and Miles (1983) have proposed radical reclassifications of economic activity in an attempt to explore the economic and social implications of the rise of the ser-vice economy. In the United States of America (USA), Browning and Single-mann (1975) and Singlemann (1979) have also broken with the traditional classifications, disaggregating services into business, distributive, personal, and collective services. Noyelle (1983) and Noyelle and Stanback (1984) have adopted a modified Singlemann classification to explore the changing structure and economic role of cities in the USA. This classification dis-tinguishes eight sectors: Agriculture, Extractive and Transformative (which includes manufacturing), Distributive Services, Corporate Activities (head offices and producer services), Nonprofit Services (Education and Health), Retail Services, Consumer Services, and Public Sector (which includes pub-lic education and health as well as government enterprises). One purpose of this classification is to pick up the rising importance of services for the US economy during the postwar period, and the implications for the structure of the US urban hierarchy.

Within the context of urban economic analysis in the United Kingdom (UK) the conventional industrial classification has tended to prevail, although at times the three-sector model has been disaggregated further (Gudgin, *et al.* 1982).

At the outset it must be recognized that the industrial disaggregation adopted to characterize the changing economic function of cities is inevitably arbitrary. The classification adopted here is in part designed to reveal how major and familiar structural changes apparent at the national level manifest themselves at the level of individual cities and groups of cities. For example, the decline of manufacturing employment since the mid-1960s and the rise of service employment are indicative of important structural changes at the national level. However, our industrial disaggregation is also designed to provide a starting-point for understanding the processes of urban structural adaptation and for exploring the evolution of the city system as a whole. In this interim study, a nine-sector disaggregation is presented (see Appendix 1), although it is anticipated that a more disaggregated approach may become appropriate as the research progresses.

The first important distinction made is between those industries producing mainly to satisfy local demand in the city and its immediate hinterland, and those industries whose markets extend to other cities, other regions, and abroad. This distinction derives, of course, from a substantial literature which emphasizes the importance of 'exports' and the 'export base' in understanding the dynamics of urban growth and decline. Three of the nine industrial groups—Groups A, G and H—are distinguished as serving mainly local markets. Groups G and H are both service sectors tied to local resident (and transient) populations, the former being mainly private service sector activity, the latter public services. Group A includes such industries as agriculture, gravel, bread, bricks, cement and construction.

Groups B, C, D, E, and F serve mainly national and international markets, and the changing economic function and performance of a particular city will depend critically on the ability of firms in these industrial groups in the city to compete successfully in these markets. Certain cities specialize in particular industries because these industries are dependent on specific, often relatively immobile, resources found only in a limited number of locations. Thus our Group B industries include fishing, mining, oil, and gas as well as industries heavily dependent on these raw materials such as coke ovens, oil refining, certain chemicals, metal manufacturing industries, pottery, sea transport, and port and inland water transport.

Groups C, D, and E comprise the heart of manufacturing industry. Group E differs from the other two manufacturing groups C and D in that it covers what might loosely be regarded as the more advanced, modern parts of manufacturing: for example, pharmaceuticals, scientific instruments, radios and electronic components, and electronic computers. By contrast, industries in Group C are traditional industries, often giving an identity to a town and emerging at a relatively early stage in Britain's industrialization. This group includes leather and clothing, furniture, the paper industries, and some of the older engineering industries. Many of the products produced in this group are also distinguished by being intensely price competitive. Group D includes, *inter alia*, mechanical engineering, vehicles, and electrical appliances, and is associated with postwar industrialization and the emergence of mass markets for many consumer durables. Mass production methods are prevalent in this group and, as with Group C, the products face intense international competition based more on product differentiation than on price.

Finally, industries in Group J are serving national and regional markets facing limited or little overseas competition. Industries in this group include, for example, gas, electricity, and water, national government service, printing and publishing, and brewing and malting. These are industries in which national demand can be supplied by one or two locations, but where there is also a regional component to demand, providing scope for regional centres to attract these industries.

City Specializations

Analysis of the proportions of total employment in each of the nine industry groups shows how individual cities are tied to particular types of activity. Clearly, a city's ability to sustain and create jobs will be considerably affected by its inherited structure of industry. Equally important to performance prospects will be the extent to which the city undergoes a structural shift out of declining sectors and towards expanding activities.

Table 3.1 shows how proportions of employment in each of the industry groups have varied in the period 1971–81. Predictably, there is a significant decline in industrial employment (Groups A–E), offset by the rise in service employment. It is, however, noteworthy that employment has declined most in the two groups (B and C) which cover the most traditional sectors. Decline is also relatively high in Group D which includes more modern, internationally competing industries. Group A industries have fared slightly less badly, reflecting their reliance more on local than national or international markets. The least bleak part of industry is the high technology sector, but even here there is a marked job loss.

The undoubted star of the service sector is Group F which comprises the more advanced, producer-oriented activities. Given that these sectors employ professional staff in far greater numbers, the location decisions of

Table 3.1. *Proportions of Employment[a] by Industry Group in Great Britain, 1971 and 1981*

Industry group	1971 (%)	1981 (%)	Change (% points)	Growth (%)
A	9.8	8.7	−1.1	−11.2
B	9.0	6.7	−2.3	−25.6
C	12.3	9.0	−3.3	−26.8
D	8.6	6.7	−1.9	−22.1
E	4.9	4.4	−0.5	−10.2
F	6.1	8.5	+2.4	+39.3
G	22.4	26.5	+4.4	+19.6
H	17.9	20.4	+2.5	+14.0
J	9.0	9.1	+0.1	+1.1

[a]Total employment in the two years is nearly the same, so that the proportions are equivalent to numbers of jobs.

firms operating in Group F industries are likely to be based far more on satisfying the residential preferences of these staff than on more conventional characteristics like transport costs or proximity to markets. This is likely to have a considerable bearing on the suitability of cities to house these industries, with those in uncongenial locations or with unattractive environments standing little chance of creating jobs.

The two largely population-related service sectors do, as expected, show employment growth, although it is notable that the same number of jobs was created in the large group H as in the very much smaller Group F. Private service jobs (Group G) have expanded at a faster rate and, because of their large weights in the total, are bound to be an important part of employment change. The number of jobs in Group J—which might be expected to be more mobile—remains virtually unchanged.

However, it is likely that the most critical determinants of a city's performance will be how well it is able to do in potentially mobile industries, rather than those like Groups A, G, and H which are much more population dependent.

Tables 3.A1–3.A9 in Appendix 2 show rankings for each industry group of the top and bottom twenty cities, by proportion of employment in the industry. Specialization in a 'wrong' industry will, on the whole, be an indication that the city's prospects are poor. Remaining specialized in a 'wrong' industry is even worse, though a city that succeeds in moving out of a 'wrong' industry must be expected to do comparatively well.

In Group A, the range in 1971 was from 20 per cent in Aberdeen, to just 2.9 per cent in Basildon. It is not, however, a group in which there are extreme specializations, as nearly all the cities have proportions of employment in Group A which are only a few percentage points off the Great Britain average. This means that, except where the city is, in fact, serving a national market, the presumption that the industry group serves mainly local markets is borne out. As such, it is not a group with great predictive value in terms of a city's likely performance.

As might be expected, Group B exhibits far greater specializations. Steel and coal towns are particularly prominent in the top twenty list, with nearly half of all employment in Dearne Valley in this group. In stark contrast, many southern towns, particularly coastal resorts, appear in the bottom twenty, with virtually no Group B industry. A further observation about Group B is that many of the towns highly specialized in this sector lost large numbers of jobs in it—50 per cent lost in, for example, Scunthorpe.

The range in Group C is also quite large, with more than 40 per cent of jobs in this group at the top, down to just 1.5 per cent at the bottom. Once again, it is a group in which specialization has diminished markedly among the top twenty. The contrast between the top and bottom twenty is to a considerable extent one between the industrial heartland of Britain and the genteel South.

Group D comprises more mixed industry, and is evidently dominated at the top end by motor vehicles. Coventry and Luton stand out as cities that have fared very badly in this group. Once again, the bottom has its complement of South Coast resorts, although it is interesting to note that the bottom twenty also includes several traditional industrial towns where this class of industry has evidently made no headway.

Group E presents a rather different picture of industry, with the South of the country and the New Towns much more represented in the top twenty. Conversely, many of the bottom twenty are older industrial towns which have evidently been unsuccessful in attracting these newer, high technology industries.

A similar pattern applies in the advanced service group F. Only Edinburgh, Chester, Blackpool, and Warrington from 'North of Watford' manage to get into the top twenty in 1981. The bottom end, however, almost reads like a catalogue of declining industrial areas. Indeed, given that a fraction of activity in this group is actually locally based, the low proportions of employment recorded by many of the bottom twenty mean that they have virtually no contribution from the national boom in this area to their employment.

Private services (Group G) are predictably high proportions of employment in resort areas and regional centres, although the range in this group is comparatively small. This is as would be expected given the link between population and service jobs. Some of the bottom twenty have quite marked increases, reflecting in part their decline elsewhere, but in part a catching up with national standards.

Group H has no outstandingly high shares, although there is evidence of educational specialization (Oxford, Bath, and Cambridge). Birmingham stands out in the bottom twenty as a large city which lacks public employment in 1971, but obtains a disproportionate increase by 1981, taking it out of the bottom twenty.

Specialization in Group J is widely dispersed geographically, as might be expected, given that it comprises sectors which serve both regional and national, but protected, markets. To some extent, the bottom twenty in this group appear to rely on neighbouring cities for these activities, suggesting a possible reason for adjacent cities having diverging job performances, on the basis of a 'central place theorem' approach to location of industry.

It is already possible from these rankings to discern some of the specializations of cities and to see how these are likely to affect employment performance. The following sections extend the analysis towards explanations of these performances.

Classes of cities

There are many different ways in which cities can be classified, including regionally, by size, and by function. A rigorous statistical approach to distinguishing between cities would be beyond the scope of this Chapter, but some insights can nevertheless be gained by examination of obvious groupings. In this section, therefore, a few such groups are analysed.

Tables 3.2 and 3.3 show, for 1971 and 1981 respectively, the proportions of total employment in each of the nine sectors of activity, as well as the gender structure of employment. In Group A industries—which mainly serve local markets—the proportion of employment does not show much variation. The one discernible pattern is that the proportion of employment in this group is inversely related to city size, with the lowest proportion occurring in London. Group H—local public services—is even more uniform, suggesting that population is the main determinant of employment in this sector.

In private local services (Group G), the proportions of employment in 1981 are also fairly constant, with the exception of the resort towns. There, it is not unreasonable to expect a markedly higher proportion, since the resident population will be swelled by tourists. More variation in local private service jobs is evident in 1971, with old industrial towns and New Towns in particular apparently lagging behind the rest of the country in this area. Elsewhere, the growth in the number of private service jobs in the London commuter towns is considerable, and can probably be attributed in large measure to the continuing exodus from the capital.

The other six sectors of activity show greater variability. Not surprisingly, the 'old industrial towns' continue to have very high shares of the older industrial sectors—Groups B and C. Conurbations, especially London, are under-represented in this area, while the buoyant 'resort' and 'Heathrow–Gatwick' areas have scarcely any Group B industry and relatively little Group C.

New Towns and the London commuter belt stand out in their degree of specialization in Group D, which embraces more of the postwar large-scale industries, especially motor vehicles. The Heathrow–Gatwick area (which of course overlaps significantly with the London commuter belt) is also well represented in Group D in 1971, though its relative specialization is much eroded by 1981. However, the area retains its pre-eminence in the high technology Group E. The medium-sized cities (those in the 200–350 thousand population range) also appear to have done comparatively well in attracting and retaining Group E industry, but the ports, the conurbations, and the older industrial towns have had rather poor records in this important group, pointing towards the location preferences of modern industry.

In Group F, the advanced business services sector, there are also big divergences. The London area generally has been most successful in attracting jobs in this growing sector. Indeed, it is the high share of employment in this group and in Group J that sharply differentiates London from the other conurbations. Medium-sized cities are also notable for the extent to which they have raised their share of Group F activity between 1971 and 1981. At the other extreme, the proportion of Group F jobs in old industrial towns is very low.

Table 3.2. *Employment by Sector based on Employment Exchanges, 1971*

	Share by industry group (%)									Share by gender (%)		Total employment
	A	B	C	D	E	F	G	H	J	Males	Females	
London	6.5	4.1	8.4	6.2	4.4	13.7	25.9	16.7	14.1	60.9	39.1	4 002 064
London commuter belt	7.5	4.9	10.5	12.4	10.5	6.7	20.4	18.4	8.7	62.2	37.8	1 328 626
Extended London area	6.7	4.3	8.9	7.8	5.9	12.0	24.5	17.2	12.7	61.2	38.8	5 330 690
Major seaports	9.7	14.0	9.8	7.2	4.7	4.9	22.9	18.5	8.3	63.0	37.0	2 009 324
Regional centres	8.4	9.7	13.9	6.6	5.0	5.2	22.6	18.9	9.9	61.4	38.6	1 937 345
Resorts	9.0	2.7	6.9	5.9	6.2	6.5	31.7	20.3	10.9	56.7	43.3	765 744
Old industrial towns	9.2	16.4	18.0	8.2	4.1	2.4	18.8	17.1	5.9	63.9	36.1	1 341 639
Heathrow–Gatwick area	7.3	2.7	9.8	10.6	10.2	8.1	22.2	18.8	10.3	60.7	39.3	804 455
New Towns	9.6	8.7	10.6	14.4	7.2	5.2	19.5	17.0	7.7	63.2	36.8	830 577
Conurbations	7.5	6.4	13.0	8.7	4.2	9.3	23.7	16.6	10.6	61.2	38.8	8 190 537
Conurbations (ex London)	8.4	8.6	17.4	11.0	4.1	5.1	21.6	16.5	7.3	61.4	38.6	4 188 473
Large cities	8.3	12.8	13.2	8.5	6.1	4.4	20.8	17.8	8.1	62.4	37.6	2 072 168
Medium cities	8.9	6.6	8.5	11.1	6.9	5.3	23.6	19.6	9.4	61.9	38.1	2 009 544
Larger towns	9.0	9.7	14.8	8.7	5.7	4.6	20.9	18.8	7.9	62.3	37.7	2 040 063
Smaller towns	10.9	12.0	10.5	8.7	5.6	4.0	21.8	18.2	8.3	62.5	37.5	1 912 211
Large towns (ex New Towns)	8.9	10.3	15.1	8.2	5.8	4.0	21.0	18.9	7.8	62.4	37.6	1 973 012
Small towns (ex New Towns)	10.7	11.9	10.5	7.6	4.8	4.0	22.8	19.2	8.5	62.2	37.8	1 527 647
Total Great Britain	9.8	9.0	12.3	8.6	4.9	6.1	22.4	17.9	9.0	62.0	38.0	21 648 457

Table 3.3. *Employment by Sector based on Employment Exchanges, 1981*

| | Share by industry group (%) | | | | | | | | | Share by gender (%) | | Total employment |
	A	B	C	D	E	F	G	H	J	Males	Females	
London	5.7	2.4	5.3	4.3	3.6	17.3	28.3	19.6	13.5	57.6	42.4	3 575 060
London commuter belt	6.8	2.8	6.9	9.2	9.1	10.3	26.5	19.4	9.0	57.1	42.9	1 383 575
Extended London area	6.0	2.5	5.7	5.7	5.1	15.4	27.8	19.6	12.3	57.5	42.5	4 958 635
Major seaports	8.3	9.8	7.5	5.6	3.9	7.6	27.1	20.6	9.6	57.5	42.5	1 788 061
Regional centres	7.1	6.9	10.0	5.0	4.6	8.2	26.5	21.2	10.5	56.6	43.4	1 853 396
Resorts	7.8	2.5	5.3	5.2	4.7	9.0	32.8	21.2	11.7	52.6	47.4	825 364
Old industrial towns	8.4	12.3	13.2	6.8	3.1	4.1	24.1	21.0	7.1	58.1	41.9	1 212 745
Heathrow–Gatwick area	6.4	1.9	6.5	7.6	9.4	11.9	27.8	18.9	9.5	56.5	43.5	874 178
New Towns	7.9	6.1	8.0	10.6	6.0	7.9	26.8	18.2	8.5	57.9	42.1	919 719
Conurbations	6.5	4.4	8.8	6.3	3.7	12.5	26.9	19.8	10.9	57.5	42.5	7 182 311
Conurbations (ex London)	7.4	6.5	12.3	8.4	3.8	7.8	25.5	20.1	8.3	57.3	42.7	3 607 251
Large cities	7.1	9.2	10.1	6.2	4.9	7.3	25.7	20.7	8.8	57.1	42.9	1 903 634
Medium cities	7.5	5.3	6.4	8.0	6.8	8.6	27.7	20.3	9.4	57.6	42.4	2 099 989
Larger towns	8.1	7.5	10.7	6.4	4.7	7.1	25.8	21.2	8.5	57.0	43.0	2 212 676
Smaller towns	9.2	8.3	7.8	7.9	4.3	6.0	27.2	20.4	8.9	56.7	43.3	1 992 918
Large towns (ex New Towns)	8.1	7.9	11.1	6.2	4.7	6.2	25.7	21.5	8.5	57.0	43.0	1 957 141
Small towns (ex New Towns)	9.2	8.3	7.7	6.8	4.4	6.1	27.2	21.3	8.9	56.8	43.2	1 549 580
Total Great Britain	8.7	6.7	9.0	6.7	4.4	8.5	26.6	20.4	9.1	57.4	42.6	21 054 359

The large and small towns are also significantly underdeveloped in these sectors, suggesting that there may well be an agglomeration effect to consider.

Finally, in Group J, there is a suggestion that the group of regional centres, together with London (though not, perhaps surprisingly, the other conurbations), have a relative specialization. Again, the old industrial towns are revealed to be comparatively unattractive locations for this sector, and smaller urban areas are also under-represented in it.

The gender composition of employment is, to a considerable extent, linked to the sectoral patterns. Thus the resorts, with disproportionate numbers of service jobs, have the highest female employment share. Ports and old industrial towns had low female shares in 1971, but by 1981 the uniformity of the female share is the most striking feature.

The Relative Growth of Britain's Cities

This section examines the growth of Britain's cities between 1971 and 1981, focusing firstly on those cities experiencing relatively rapid growth and secondly on the group of cities stagnating or declining. The analysis then proceeds to explore some of the factors associated with differences in growth rates across cities.

The rapidly growing cities

Table 3.4 shows the twenty fastest-growing cities (of the total 104 cities in our analysis) ranked by their rate of total employment growth over the period 1971–81. Several features of the table deserve comment. Firstly, eighteen of the twenty cities are located in the southern half of the country, and the majority of these enjoy good access to London and indeed are within commuting range of the capital. Five of the cities are part of the latest phase of the postwar New and Expanded Towns programmes: Milton Keynes, Peterborough, Crawley, Basildon, and Basingstoke. Female employment grew more rapidly than male employment, but this broadly reflects national trends. On average, employment growth in the top twenty cities exceeded national employment growth by about 25 per cent over the decade.

To explore whether fast-growing cities showed evidence of industrial specialization in terms of our nine-sector industrial classification, location quotients were computed by sector for each of the cities for the beginning of the period, 1971 (see Table 3.5). The average of the location quotients for the top twenty rapidly growing cities reveals a strikingly low relative proportion of activity in industrial Groups B and C. It will be recalled that these groups cover industries serving national and international markets and include, *inter*

Table 3.4. *The Twenty Fastest-growing Cities in Great Britain, 1971–1981* (Total 1971 Employment = 100)

City	Index as at 1981		
	Total	Males	Females
Milton Keynes	189.7	177.9	210.1
Aberdeen	137.4	139.2	134.7
Basingstoke	136.3	121.0	162.3
Cheltenham	133.8	127.8	142.8
Basildon	127.7	120.4	139.6
Reading	124.8	120.1	131.5
Aldershot	121.3	108.9	140.0
Northampton	119.5	110.6	132.3
Peterborough	118.8	107.6	142.3
Eastbourne	118.4	112.0	125.8
Maidstone	117.1	110.2	128.3
High Wycombe	116.0	104.8	134.0
Cambridge	115.5	109.1	124.9
Bedford	114.4	102.0	134.9
Exeter	114.2	104.9	128.1
Harrogate	113.6	105.6	124.1
Hertford	113.5	99.7	135.7
Colchester	112.6	103.5	127.1
Southend	111.0	102.5	120.2
Crawley	110.7	102.7	123.3
Average of top twenty cities	123.3	114.5	137.2
Great Britain	97.3	90.0	109.0
Average growth of top twenty cities relative to GB growth	126.7	127.2	125.8

alia, the old and large declining industries of shipbuilding, mining, and iron and steel established in the early stages of industrialization, as well as the capital-intensive sectors of oil, general chemicals, and tobacco. However, these cities are relatively well endowed with those industries associated more closely with the postwar wave of industrialization, Groups D and E. Group E, which includes a number of the more modern or 'high technology' industries, is particularly well represented, although only eleven of the twenty cities had a location quotient greater than 100 in this group. This compares with a total of fourteen cities with a location quotient greater than 100 for Group H industries, which include mainly the major public services of education, health, and local government services.

The pattern of specialization for individual cities in this fast-growing group is also very variable, although for every fast-growing city the location quotient for Group B industries was significantly below 100. However, Hertford, Basingstoke, Cheltenham, and Basildon, for example, have location

Table 3.5. *Location Quotients for each of the Nine Industrial Groups for the Twenty Fastest-growing Cities as at 1971 and 1981*

City	Industry group: 1971									Industry group: 1981								
	A	B	C	D	E	F	G	H	J	A	B	C	D	E	F	G	H	J
Milton Keynes	140	36	159	120	102	61	71	100	118	92	30	119	88	123	98	100	124	88
Aberdeen	204	70	81	24	10	74	126	121	76	147	216	59	33	27	85	113	97	72
Basingstoke	72	3	76	172	229	78	106	90	141	84	19	62	145	216	68	120	95	103
Cheltenham	121	20	49	72	243	136	121	101	96	107	61	50	112	202	135	92	80	146
Basildon	30	13	107	330	337	41	68	79	66	50	19	84	416	200	66	86	78	60
Reading	78	17	66	87	143	144	104	133	137	71	31	51	61	202	160	112	92	133
Aldershot	60	14	31	92	180	313	100	74	196	63	30	22	116	166	221	103	74	154
Northampton	76	16	120	199	51	107	102	110	89	91	33	124	163	46	115	108	102	69
Peterborough	153	31	16	335	41	52	101	83	96	118	37	48	351	27	76	101	74	104
Eastbourne	138	9	24	38	34	75	177	138	97	85	34	20	87	41	59	144	139	103
Maidstone	133	52	144	59	18	77	95	144	74	133	36	142	40	25	95	97	133	95
High Wycombe	67	19	198	183	120	90	84	84	71	71	28	170	176	118	119	99	83	66
Cambridge	108	30	34	49	167	125	106	168	88	100	30	32	90	132	102	92	164	86
Bedford	160	71	52	39	155	156	91	120	104	129	66	58	51	171	107	92	124	106
Exeter	144	38	27	16	33	100	126	169	129	103	24	33	34	41	93	111	165	124
Harrogate	118	24	49	60	20	159	127	129	141	136	18	43	52	32	74	124	121	157
Hertford	63	49	104	85	451	98	83	102	44	55	55	87	58	443	94	92	115	48
Colchester	161	47	27	79	78	64	107	130	166	114	32	41	152	107	75	112	109	121
Southend	79	26	80	99	78	116	127	115	131	80	25	81	119	71	142	113	86	145
Crawley	76	43	37	152	229	228	90	101	82	77	25	34	113	218	231	96	90	87
Average for the top twenty cities	109	32	75	114	135	115	106	115	107	95	42	68	123	130	111	105	107	103
Number of cities with LQ >100	11	0	6	7	11	10	12	14	9	9	1	4	10	12	9	12	10	11

Note: LQ, location quotient.

quotients exceeding 200 for Group E industries, whereas Aberdeen, Peter-borough, Eastbourne, Maidstone, Exeter, and Harrogate each have quotients below 50 for this group.

If we examine the location quotients for 1981 the picture is broadly the same, although there is some evidence of increased specialization in Group D and decreased specialization in Group A (see Table 3.5).

The declining cities

The bottom twenty cities losing employment are primarily located in the North of England, Scotland, and the Midlands (Table 3.6). Four of the cities—Liverpool–Birkenhead, Glasgow, Tyneside, and Birmingham—are major conurbations where significant employment loss is widely recognized. The other cities are typical of Britain's old and long-established manufacturing centres.

Table 3.6. *The Bottom Twenty Declining Cities in Great Britain, 1971–1981* (Total 1971 Employment = 100)

City	Index as at 1981		
	Total	Males	Females
Sunderland	68.7	56.1	87.0
Liverpool–Birkenhead	76.3	70.1	86.1
Scunthorpe	77.1	66.2	106.2
Dearne Valley	79.0	97.5	79.0
Airdrie–Coatbridge	79.4	66.7	101.7
Huddersfield	80.8	77.9	85.8
Kettering	82.2	69.9	103.2
Thames Estuary	82.6	74.6	100.4
Coventry–Nuneaton	82.7	73.8	100.5
Glasgow	83.6	76.6	94.1
Newport (Wales)	84.5	73.9	107.6
Tyneside	84.9	78.5	95.3
Blackburn	85.2	80.2	93.1
Birmingham	86.1	81.0	95.0
Hartlepool	86.3	79.4	98.0
Sheffield	86.4	78.9	99.6
Greenock	86.7	78.4	101.5
The Potteries	87.9	84.1	93.7
Cardiff	88.6	79.0	104.1
Motherwell	88.6	84.5	98.7
Average of bottom twenty cities	82.9	76.4	96.5
Great Britain	97.3	90	109
Average of bottom twenty cities relative to GB	85.2	84.9	88.5

Table 3.7. *Location Quotients for each of the Nine Industrial Groups for the Bottom Twenty Declining Cities as at 1971 and 1981*

City	Industry group: 1971									Industry group: 1981								
	A	B	C	D	E	F	G	H	J	A	B	C	D	E	F	G	H	J
Sunderland	54	196	81	88	176	38	116	92	69	56	118	71	73	75	53	143	111	84
Liverpool–Birkenhead	86	129	95	127	86	88	101	101	78	84	133	116	124	73	98	97	101	82
Scunthorpe	180	501	24	5	—	33	72	71	37	174	461	36	24	18	46	91	78	48
Dearne Valley	59	519	56	115	—	15	48	86	38	115	470	116	82	2	45	65	94	25
Airdrie–Coatbridge	125	160	96	107	300	21	88	74	39	139	176	78	64	125	52	94	122	54
Huddersfield	67	30	280	149	98	39	71	87	53	71	46	204	212	104	73	85	93	64
Kettering	88	353	150	53	8	44	69	80	39	93	267	150	79	32	54	93	87	74
Thames Estuary	116	152	175	70	47	31	87	94	76	120	158	131	88	46	48	101	101	86
Coventry–Nuneaton	41	56	55	369	353	38	71	73	44	51	82	73	288	291	59	81	99	54
Glasgow	119	103	94	99	55	87	111	105	82	116	108	91	70	59	97	101	117	90
Newport (Wales)	80	329	30	104	6	46	92	107	80	84	294	49	106	39	56	92	107	97
Tyneside	105	147	76	69	125	64	102	97	122	92	154	77	73	93	78	101	98	136
Blackburn	56	11	260	120	145	41	76	84	108	55	19	338	93	96	41	77	100	97
Birmingham	80	112	157	203	69	70	81	76	66	74	116	157	210	84	87	84	87	74
Hartlepool	150	257	82	23	231	30	74	88	52	156	176	128	49	177	29	83	91	98
Sheffield	90	241	140	66	20	59	88	96	59	89	207	136	79	39	77	92	109	68
Greenock	82	346	88	50	110	33	77	84	69	54	363	49	33	250	35	84	100	85
The Potteries	89	350	78	71	80	38	73	91	57	100	369	103	93	84	46	83	78	62
Cardiff	98	116	56	39	25	97	108	137	152	93	79	58	43	27	102	109	118	171
Motherwell	139	356	42	78	101	25	73	93	38	141	322	30	151	134	34	75	97	51
Average of the bottom twenty cities	95	223	106	100	102	47	84	91	68	98	197	110	102	92	61	92	99	80
Number of cities with LQ >100	7	17	6	8	8	—	5	4	3	7	16	10	6	5	1	5	8	2

Note: LQ, location quotient.

Table 3.7 reveals very clearly the relative specialization of declining cities on Group B industries. The location quotient averages over 200, and reaches 500 in Scunthorpe, which specializes in iron and steel.

Another distinguishing characteristic of these cities is indicated by the low location quotients of the service sector—Groups F, G, H, and J. Not only are they each well below 100, but they contrast strikingly with the location quotients for these industries in the fast-growing cities which are all above 100. In other words, it is the below-average share of the labour-intensive and more rapidly growing service sectors which (along with an above-average share of old-established declining industries) is associated with the poor employment performance of this group of cities. Table 3.7, which gives location quotients for 1981, shows much the same picture as for 1971.

Preliminary analysis of factors associated with differences in the growth of employment across cities

Our starting-point for examining factors associated with the differential employment performance of cities was to undertake a simple shift–share analysis of the fastest- and slowest-growing group of cities. As is well known, this type of analysis serves merely to distinguish the contribution of different components of change (national, structural, and differential). The split between structural and differential contributions depends very much on the degree of industrial disaggregation, so the results presented in Table 3.8 should be treated with caution, bas they are on only a nine-sector dis-aggregation. Nevertheless, what is clear from Table 3.8 is the relative importance of the differential component in its contribution to actual employment. This component, it will be recalled, is the difference between actual employment and employment expected, where the latter is the change in total employment if each of the nine groups grew at the national rate. A positive differential suggests 'city-specific' factors raising the growth rate above that occurring nationally, and a negative differential the reverse. What emerges clearly from Table 3.8 is a consistently positive differential for fast-growing cities and a consistently negative differential for slower-growing cities. By contrast, the contribution of industrial structure is both positive and negative for the fast-growing cities, although it is negative (and relatively small) in the majority of declining cities.

Although the shift–share analysis in Table 3.8 provides no explanation of differences in employment growth across cities, it does nevertheless provide a useful starting-point for a more formal statistical analysis. What is clear from the shift–share analysis is that, even after allowing for major differences in industrial structure, substantial differential growth of employment effects remain to be explained. In the following statistical analysis, therefore, the dependent variable is the 'differential' employment growth for each of the

Table 3.8. *The Percentage Contribution of the Differential Growth and Structural Components to Actual Employment in each of the Twenty Fastest-growing and Twenty Slowest-growing Cities, 1971–1981*

Twenty fastest-growing cities

City	Differential growth contribution	Structural contribution
Milton Keynes	+50.4	−1.7
Aberdeen	+26.9	+2.2
Basingstoke	+28.2	+0.7
Cheltenham	+23.8	+3.5
Basildon	+28.5	−4.7
Reading	+17.8	+4.3
Aldershot	+12.3	+7.4
Northampton	+18.1	+0.4
Peterborough	+19.5	−1.5
Eastbourne	+10.1	+7.7
Maidstone	+16.4	+0.7
High Wycombe	+19.2	−3.2
Cambridge	+9.7	+6.0
Bedford	+11.6	+3.5
Exeter	+8.0	+6.9
Harrogate	+7.8	+6.6
Hertford	+15.0	−0.6
Colchester	+10.4	+3.3
Southend	+8.2	+4.2
Crawley	+8.0	+4.2

Twenty slowest-growing cities

City	Differential growth contribution	Structural contribution
Sunderland	−38.5	−2.9
Liverpool–Birkenhead	−26.1	−1.3
Scunthorpe	−15.3	−10.9
Dearne Valley	−8.2	−15.1
Airdrie–Coatbridge	−15.6	−6.8
Huddersfield	−10.9	−9.6
Kettering	−7.6	−10.9
Thames Estuary	−11.5	−6.3
Coventry–Nuneaton	−10.1	−7.6
Glasgow	−17.0	+0.7
Newport	−10.8	−4.1
Tyneside	−14.0	−0.7
Blackburn	−7.4	−6.7
Birmingham	−6.6	−6.5
Hartlepool	−6.3	−6.4
Sheffield	−7.1	−5.3
Greenock	−4.2	−8.1
The Potteries	−2.8	−7.9
Cardiff	−14.0	+4.3
Motherwell	−2.4	−7.5

104 cities in the analysis. This differential growth is the actual employment growth in each city minus the employment growth that would be expected if each industry in the city grew at the same rate as its national counterparts (that is, the 'expected' employment). The implicit assumption here is that national factors working on employment change in the city are captured by 'expected employment', and that all other factors influencing city employment change are city-differentiated effects of both a policy and a non-policy kind. Such effects working nationally on all cities would include the impact on employment arising from national macro-economic policies, or from technological changes common to an industry irrespective of its city location. City-differentiated effects would include urban policies, peripherality, and agglomeration economies and diseconomies associated with city size for example. These latter factors (and other city-specific factors) are therefore the variables which account for or explain city-differentiated employment effects.

In the following analysis, only a limited number of explanatory variables are included. The size of the city is included to pick up employment change resulting from agglomeration economies and diseconomies. In the recent period, the larger cities and conurbations have been experiencing a relative loss of employment compared with smaller cities. *Inter alia*, this may reflect possible agglomeration diseconomies or the increased potential mobility of many industries in choosing their location. A priori, it would be expected that the coefficient on city size would be negative.

New Town policy has been an important factor behind the rapid employment growth of certain cities as revealed in Table 3.8. To pick up the impact of this policy, a dummy variable was used, taking a value of 1 for those cities with New or Expanded Town status and 0 for all other cities.

Finally, the third element introduced in the preliminary regressions was a dummy to capture the poor employment performance associated with cities in the peripheral regions of the country. Typically, in the absence of regional policy the employment performance of these regions is relatively poor. Regional policy improves the relatively poor performance of manufacturing industry in particular, and it is difficult therefore to say a priori whether the coefficient on the peripheral city dummy variable will be positive or negative.

A linear equation was estimated by ordinary least squares (OLS) for 100 cities. The variables in the regression were:

$$\frac{A-E}{A} = \frac{\text{Actual} - \text{Expected Employment}}{\text{Actual Employment}} \times 100 \quad \text{for the period 1971–81}$$

CSZ = city size (measured in terms of total employment in 1971);
NT = a dummy variable for New Town and Expanded Town status; and

REG = a dummy variable for cities in the North, Scotland, the North West, and Wales.

The estimated equation is

$$\frac{A-E}{A} = \underset{(4.8)}{62.956} - \underset{(-4.5)}{5.157} \log \text{CSZ} - \underset{(-3.2)}{6.411} \text{REG} + \underset{(4.3)}{14.027} \text{NT} \qquad (3.1)$$

$$R^2 = 0.36 \quad \text{DW} = 1.62 \quad F = 18.36$$

Equation (3.1) is based on 100 observations for 1971–81; the T-ratios are shown in parentheses. The equation shows that each of the variables—city size (CSZ), New Towns (NT), and 'peripherality' (REG)—is statistically significant. As anticipated, the sign on the New Town dummy is positive, indicating the contribution to employment growth of New Town designation. The coefficient on city size is negative, suggesting perhaps agglomeration diseconomies and the presence of other factors favouring the growth of employment in the smaller cities.

A variant of equation (3.1) is to separate out specifically the 'national' and 'structural' components of the 'expected' variable (or differential effect). This is achieved by formulating a new dependent variable which takes out the 'national' effect from actual employment change in each city: $A - N$. The structural effect is now included in the equation as an 'explanatory' variable S/A:

$$\frac{A-N}{A} = \underset{(4.9)}{0.648} - \underset{(-4.6)}{0.053} \log \text{CSZ} - \underset{(-3.4)}{0.662} \text{REG} + \underset{(4.4)}{0.145} \text{NT} + \underset{(6.9)}{1.308} S/A \qquad (3.2)$$

$$R^2 = 0.50 \quad \text{DW} = 1.66 \quad F = 23.55$$

To pursue the question of the importance of different sectors in contributing to the differential employment performance of cities a further series of regressions were carried out. In these regressions the dependent variable is the percentage change in total employment in the period 1971–81. City size, New Towns and peripherality remain in the regression but, in addition, the share of employment in 1971 in selected sectors is included. Equation (3.3) shows the OLS result with sectors B, C, D, and E (BCDE) as a group (comprising therefore mainly mining and manufacturing, and subject to competition from imports):

$$\% \triangle E = \underset{(4.68)}{76.3} - \underset{(-5.09)}{0.488} \text{BCDE} - \underset{(-3.53)}{4.94} \log \text{CSZ} + \frac{20.0\text{NT}}{(4.95)} - \underset{(-3.09)}{7.51\text{REG}} \qquad (3.3)$$

$$R^2 = 0.42 \quad \text{DW} = 1.62 \quad F = 17.3$$

Each of the variables in equation (3.3) is statistically significant. The coefficient on the combined share of the four sectors BCDE is negative, indicating that cities with an above-average share of these sectors in 1971 grew relatively slowly. The sign of the other coefficients confirms the findings of equations (3.1) and (3.2).

The second industrial group—the share of sectors A, G, H, and J (the sectors which are considered mainly to serve local and regional markets, and which are not exposed to import penetration)—is found to be positively associated with employment growth:

$$\% \triangle E = \frac{22.0}{(1.20)} + \frac{0.517}{(4.68)} \text{AGHJ} - \frac{4.40}{(-3.07)} \log \text{CSZ} + \frac{20.6\text{NT}}{(5.0)} - \frac{7.96\text{REG}}{(-3.20)} \quad (3.4)$$

$$R^2 = 0.40 \quad \text{DW} = 1.64 \quad F = 16.0$$

Finally, in these preliminary investigations each of the nine sectors in the analysis was included separately in a regression to test for its individual association with city employment performance. The results of these regressions are shown in Table 3.9.

Table 3.9. *Results of Regression Analysis to Test for the Association of each of the Nine Industrial Sectors with City Employment Performance*

Regression including sector	Constant	Coefficient on				R^2	DW
		Sector	CSZ	NT	REG		
A	46.4	0.69	−4.5[a]	16.8[a]	−6.6[a]	0.28	1.46
B	72.9[a]	−0.58[a]	−5.8[a]	16.5[a]	−5.4[a]	0.43	1.72
C	61.2[a]	−0.16	−5.64[a]	16.8[a]	−6.1[a]	0.27	1.55
D	62.1[a]	−0.20	−5.1[a]	18.8[a]	−6.7[a]	0.27	1.50
E	59.7[a]	0.29	−5.2[a]	17.2[a]	−6.0[a]	0.27	1.57
F	73.7[a]	1.86[a]	−7.1[a]	15.5[a]	−4.7[a]	0.37	1.52
G	43.8[a]	0.68[a]	−5.0[a]	19.6[a]	−7.5[a]	0.33	1.60
H	33.4	1.18[a]	−4.69[a]	20.4[a]	−7.0[a]	0.36	1.71
J	54.6[a]	1.10[a]	5.5[a]	17.7[a]	−6.0[a]	0.33	1.6

[a] Statistically significant at 5% level of significance.

This analysis shows that sectors B, F, G, H, and J are statistically significant (as are the other variables in every equation) when included separately in each regression. The coefficient on B is negative while those on sectors F, G, H, and J are positive. One particular observation on the size of the coefficients is the relatively large coefficient on sector F, which suggests that a high concentration of employment in that sector is associated with more rapid employment growth.

Conclusions

Using a new nine-sector classification of economic activity by employment, we have investigated the growth and specialization of Britain's hundred largest cities over the period 1971–81. A considerable range of specialization is shown to exist, with the northern cities dominant in the traditional manufacturing and resource-based industries of Groups B and C, and the southern cities revealing a stronger representation in the more modern industries of Groups E and F. As might be expected, there is little variability across cities in Groups A and H which are activities closely related to the cities' population. Private services (Group G) show predictably high proportions of employment in resort cities and regional centres, although, being tied to population, the range of the share of this group is comparatively small. Specialization in Group J is widely dispersed geographically given that it comprises sectors serving local, regional, and national markets.

Eighteen of the twenty fastest-growing cities are in the southern half of the country, and an analysis of location quotients reveals a strikingly low relative proportion of economic activity in Groups B and C. By contrast, the twenty slowest-growing cities are primarily located in the North of the country and show a heavy specialization in Group B industries and a relative absence of activity in Groups F and J.

The results of the regression analysis were encouraging and confirmed a priori expectations with respect to the importance of city size, New Town designation, industrial structure, and region. Each was found to be statistically significant in 'explaining' the employment performance of cities. The importance of particular industrial groups—notably Group B and the service sector groups—in understanding variations in city growth was also suggested.

References

Browning, H. C. and Singlemann, J. (1975), *The Emergence of a Service Society*, Springfield, Va.: National Technical Information Service.

Gershuny, J. I. (1983), *Social Innovation and the Division of Labour*, Oxford: OUP.

Gershuny, J. I. and Miles, I. D. (1983), *The New Service Economy*, London: Frances Pinter.

Gudgin, G., Moore, B. C., and Rhodes, J. (1982), *Employment Problems in the Cities and Regions of the UK*, Aldershot: Gower Publishing.

Noyelle, T. J. (1983), 'The Rise of Advanced Services: Some Implications for Economic Development in US Cities', *Journal of the American Planning Association*, 49(3), 280–9.

Noyelle, T. J. and Stanback, T. M., Jr. (1984), *The Economic Transformation of Cities*, Totowa: Rowman and Allenheld.

Office of Population Censuses and Surveys (1985), *Key Statistics for Urban Areas*, London: HMSO.

Singlemann, J. (1979), *From Agriculture to Services*, Beverley Hills: Sage.

Appendix 1: Industrial Groups used in the Analysis

The nine industrial groups used in the analysis of city employment were constructed from the Minimum List Headings (MLHs) of the 1968 Standard

Industrial Classification. The MLHs included in each group are shown below by name and MLH number.

Group A

Agriculture and horticulture (001); Forestry (002); Chalk, clay, sand, and gravel extraction (103); Bread and flour confectionery (212); Bacon curing, meat and fish products (214); Industrial plant and steelwork (341); Bricks, fireclay, and refractory goods (461); Cement (464); Construction (500).

Group B

Fishing (003); Coal mining (101); Stone and slate quarrying and mining (102); Petroleum and natural gas (104); Other mining and quarrying (109); Grain milling (211); Milk and milk products (215); Sugar (216); Fruit and vegetable products (218); Vegetable and animal oils and fats (221); Other drink industries (239); Coke ovens and manufactured fuel (261); Mineral oil refining (262); Lubricating oils and greases (263); General chemicals (271); Iron and steel (311); Steel tubes (312); Iron castings (313); Aluminium and aluminium alloys (321); Copper, brass and copper alloys (322); Other base metals (323); Shipbuilding and marine engineering (370); Pottery (462); Abrasives and building materials n.e.s (469); Timber (471); Sea transport (105); Port and inland water transport (706).

Group C

Biscuits (213); Cocoa, chocolate and sugar confectionery (217); Animal and poultry foods (219); Food industries n.e.s (229); Tobacco (240); Paint (274); Soups and detergents (275); Synthetic resins and plastic materials and synthetic rubber (276); Dyestuffs and pigments (277); Fertilizers (278); Ordnance and small arms (342); Watches and clocks (352); Locomotives and railway track equipment (384); Railway carriages and wagons and trams (385); Engineers' small tools and gauges (390); Hand tools and implements (391); Cutlery, spoons, and forks (392); Bolts, nuts, and screws (393); Wire (394); Cans and metal boxes (395); Jewellery and precious metals (396); Metals n.e.s. (399); Spinning and doubling (412); Weaving (413); Woollen and worsted (414); Jute (415); Rope, twine, and net (416); Hosiery (417); Lace (418); Carpets (419); Narrow fabrics (421); Made-up textiles (422); Textile finishing (423); Other textile (429); Leather (413), (432); Fur (433); Weatherproof outwear (441); Men's and boys' tailored outwear (442); Women's and girls' tailored outwear (443); Overalls and men's shirts etc. (444); Dresses, lingerie (445); Hats, caps, millinery (446); Dress industries n.e.s. (449); Footwear (450); Glass (463); Furniture and upholstery (472); Bedding (473); Shops and office fitting (474); Wood containers and baskets

(475); Miscellaneous wood (479); Paper and board (481); Packaging (482); Stationery (483); Paper n.e.s. (484); Rubber (491); Linoleum, plastics, floor covering (492); Brushes, brooms (493); Miscellaneous stationers' goods (495); Miscellaneous manufacturing (499).

Group D

Toilet preparations (273); Other chemical (279); Agricultural machinery (331); Pumps, valves, etc. (333); Industrial engines (334); Textile machinery (335); Construction and earth-moving equipment (336); Mechanical handling (337); Other machinery (339); Other mechanical engineering (349); Insulated wires and cables (362); Telegraph and telephone apparatus (363); Broadcast receiving etc. (365); Electrical appliances (368); Other electrical (369); Wheeled tractors (380); Motor vehicles (381); Motorcycles, cycles (382); Man-made fibres (411); Toys, games, etc. (494); Plastic products n.e.s. (496).

Group E

Pharmaceutical chemicals (272); Metal-working machine tools (332); Office machinery (338); Photographic and document copying (351); Surgical instruments and appliances (353); Scientific and industrial instruments (354); Electrical machinery (361); Radio and electrical components (364); Electronic computers (366); Radio, radar, etc. (367); Aerospace equipment (383).

Group F

Air transport (707); Insurance (860); Banking and bill discounting (861); Other financial institutions (862); Property owning and managing (863); Advertising and market research (864); Other business services (865); Central offices (866); Accountancy services (871); Research and development (876); Other professional services (879).

Group G

Road haulage (703, 704); Miscellaneous transport (709); Wholesaling (810, 811, 812); Retailing (820, 821, 831, 832); Legal services (873); Religious organizations (875); Sport etc. (882); Betting (883); Hotels (884); Restaurants (895); Public houses (886); Clubs (887); Catering (888); Hairdressing (889); Private services (891); Laundries (892); Dry cleaning (893); Motor repairs (894); Boots and shoes repair (895); Other services (899).

Group H

Railways (701); Road passenger (702); Education (872); Medical (874); Local government service (906).

Group J

Brewing and malting (231); Soft drinks (232); Printing and publishing of newspapers etc. (485, 486, 489); Gas (601); Electricity (602); Water (603); Post (708); Cinemas, theatres, etc. (881); National government service (901).

Appendix 2: The Rankings for Each Industry Group of the Top and Bottom Twenty Cities, by Proportion of Employment in the Industry

Table 3.A1. *Employment in the Top and Bottom Twenty Cities in Industry Group A as a Percentage of Total Employment in 1971 and 1981*

Top twenty

Rank	City (1971)	% 1971	City (1981)	% 1981	Rank 1971
1	Aberdeen	20.0	Cannock	15.7	13
2	Scunthorpe	17.6	Grimsby	15.7	20
3	Darlington	16.8	Scunthorpe	15.1	2
4	Colchester	15.8	Hartlepool	13.6	8
5	Bedford	15.7	Darlington	13.2	3
6	Peterborough	15.0	Carlisle	13.2	12
7	Teesside	15.0	Ipswich	12.9	18
8	Hartlepool	14.7	Aberdeen	12.8	1
9	Lincoln	14.6	Motherwell and Wishaw	12.3	16
10	Exeter	14.1	Lincoln	12.1	9
11	Chester-le-Street	14.1	Airdrie and Coatbridge	12.1	—
12	Carlisle	14.0	Harrogate	11.8	—
13	Cannock	13.9	Maidstone	11.6	—
14	Milton Keynes	13.7	Burton upon Trent	11.4	—
15	Widnes–Runcorn	13.7	Bedford	11.2	5
16	Motherwell and Wishaw	13.6	Morecambe	11.0	—
17	Eastbourne	13.5	Thanet	10.9	—
18	Ipswich	13.3	Barnsley	10.9	—
19	Warrington	13.2	Telford	10.8	—
20	Grimsby	13.1	Hull	10.8	—

Great Britain %   9.8 (1971)   8.7 (1981)

Bottom twenty

Rank	City (1971)	% 1971	City (1981)	% 1981	Rank 1971
20	Bournemouth	6.6	Huddersfield	6.2	18
19	St Helens	6.6	Mansfield	6.1	—
18	Huddersfield	6.6	Luton	6.0	4
17	London	6.5	Gloucester	5.8	—
16	Portsmouth	6.4	London	5.7	17
15	Hertford	6.2	Nottingham	5.6	14
14	Nottingham	6.2	Aldershot	5.5	13
13	Aldershot	5.9	Blackpool	5.5	—
12	St Albans	5.9	Slough	5.3	—
11	Hemel Hempstead	5.9	Hemel Hempstead	5.2	11
10	West Yorkshire	5.9	Halifax	5.2	7
9	Dearne Valley	5.8	Sunderland	4.9	5
8	Blackburn	5.5	Hertford	4.8	15
7	Halifax	5.5	Blackburn	4.8	8
6	Harlow and Bishop's Stortford	5.3	Portsmouth	4.7	16
5	Sunderland	5.3	Greenock	4.7	—
4	Luton	5.2	Basildon	4.4	1
3	Burnley	4.9	Coventry and Nuneaton	4.4	2
2	Coventry and Nuneaton	4.0	Burnley	4.3	3
1	Basildon	2.9	Torbay	4.0	—

Table 3.A2. *Employment in the Top and Bottom Twenty Cities in Industry Group B as a Percentage of Total Employment in 1971 and 1981*

Top twenty

Rank	City	% 1971	% 1981	Rank 1971
1	Dearne Valley	46.7	31.5	1
2	Scunthorpe	45.1	30.9	2
3	Mansfield	34.9	27.6	3
4	Widnes Runcorn	33.4	26.9	11
5	Teesside	32.6	25.7	5
6	Motherwell and Wishaw	32.0	25.0	8
7	Kettering	31.8	24.7	9
8	The Potteries	31.5	24.3	18
9	Greenock	30.9	22.3	15
10	Newport	29.6	21.9	12
11	Falkirk	28.8	21.6	6
12	Chesterfield	27.6	21.6	10
13	Cannock	27.6	19.7	13
14	Chester-le-Street	24.0	18.5	7
15	Barnsley	23.8	17.9	—
16	Hartlepool	23.1	17.3	19
17	Sheffield	21.7	15.2	—
18	Doncaster	20.7	14.5	—
19	Grimsby	20.1	14.4	17
20	Chatham	19.4	13.9	—

Great Britain %   0.9 (1971)   6.7 (1981)

Bottom twenty

Rank	City	% 1971	City	% 1981	Rank 1971
20	Brighton	2.1	Cambridge	2.0	—
19	York	2.1	Morecambe	2.0	—
18	St Albans	2.0	High Wycombe	1.9	—
17	Leicester	1.9	Hemel Hempstead	1.8	—
16	Cheltenham	1.8	Swindon	1.8	—
15	High Wycombe	1.7	Southend	1.7	—
14	Luton	1.6	Crawley	1.7	—
13	Reading	1.5	Exeter	1.6	20
12	Torbay	1.5	Brighton	1.5	14
11	Northampton	1.4	Luton	1.4	1
10	Hamilton and E. Kilbride	1.4	Basingstoke	1.3	—
9	Aldershot	1.3	St Albans	1.3	—
8	Southport	1.3	Basildon	1.3	7
7	Basildon	1.2	Blackburn	1.3	5
6	Oxford	1.0	Hamilton and E. Kilbride	1.3	10
5	Blackburn	1.0	Harrogate	1.2	—
4	Chelmsford	0.9	Torbay	1.0	12
3	Eastbourne	0.8	Oxford	0.7	6
2	Bath	0.5	Chelmsford	0.5	4
1	Basingstoke	0.3	Southport	0.5	8

Table 3.A3. *Employment in the Top and Bottom Twenty Cities in Industry Group C as a Percentage of Total Employment in 1971 and 1981*

Top twenty

Rank	(1971)	% 1971	(1981)	% 1981	Rank 1971
1	St Helens	43.4	Halifax	31.3	2
2	Halifax	39.4	Blackburn	30.4	4
3	Huddersfield	34.4	St Helens	26.1	1
4	Blackburn	32.0	Burnley	25.4	5
5	Burnley	30.7	Mansfield	23.2	9
6	Leicester	25.6	Leicester	20.4	6
7	W. Yorkshire	25.6	Huddersfield	18.4	3
8	High Wycombe	24.4	York	18.2	18
9	Mansfield	22.6	Burton on Trent	16.3	14
10	Thames Estuary	21.5	W. Yorkshire	15.6	7
11	Dundee	21.5	Nottingham	15.4	12
12	Nottingham	21.2	High Wycombe	15.3	8
13	Greater Manchester	20.9	Barnsley	14.2	—
14	Burton upon Trent	20.8	Birmingham	14.1	16
15	Milton Keynes	19.6	Greater Manchester	13.7	13
16	Birmingham	19.3	Wakefield and Dewsbury	13.6	—
17	Warrington	19.1	Kettering	13.5	19
18	York	18.9	Carlisle	13.0	—
19	Kettering	18.5	Maidstone	12.8	—
20	Wigan	18.1	Doncaster	12.4	—

Bottom twenty

Rank	(1971)	% 1971	(1981)	% 1981	Rank 1971
20	Southampton	5.2	Peterborough	4.3	4
19	Brighton	5.1	Bournemouth	4.2	—
18	Chatham	5.1	Portsmouth	4.1	—
17	Motherwell and Wishaw	5.1	Harrogate	3.9	—
16	Crawley	4.6	Edinburgh	3.8	
15	Cambridge	4.2	Colchester	3.7	10
14	Widnes–Runcorn	4.1	Chatham	3.5	18
13	Aldershot	3.8	Bath	3.3	8
12	Gloucester	3.7	Scunthorpe	3.2	5
11	Newport	3.7	Crawley	3.1	16
10	Colchester	3.3	Exeter	3.0	9
9	Exeter	3.3	Cambridge	2.9	15
8	Bath	3.3	Gloucester	2.7	12
7	Hastings	3.1	Motherwell and Wishaw	2.7	
6	Eastbourne	2.9	Chester	2.7	17
5	Scunthorpe	2.9	Aldershot	2.4	13
4	Peterborough	2.0	Chelmsford	2.0	2
3	Oxford	1.9	Eastbourne	1.8	6
2	Chelmsford	1.7	Oxford	1.6	3
1	Torbay	1.5	Torbay	1.4	1

Great Britain %   12.3 (1971)   9.0 (1981)

Table 3.A4. *Employment in the Top and Bottom Twenty Cities in Industry Group D as a Percentage of Total Employment in 1971 and 1981*

Top twenty

Rank	City	% 1971	City	% 1981	Rank 1971
1	Luton	40.0	Basildon	27.9	4
2	Coventry and Nuneaton	31.7	Luton	26.3	1
3	Peterborough	28.8	Peterborough	23.5	3
4	Basildon	28.4	Coventry and Nuneaton	19.3	2
5	Telford	26.4	Telford	15.7	5
6	Oxford	19.8	Cannock	15.6	13
7	Lincoln	19.8	Slough	15.4	12
8	Swindon	19.7	Chelmsford	14.9	—
9	Preston	17.6	Huddersfield	14.2	—
10	Birmingham	17.5	Birmingham	14.1	10
11	Northampton	17.1	Oxford	12.8	6
12	Slough	16.3	Darlington	12.6	—
13	Cannock	16.2	Lincoln	11.9	7
14	High Wycombe	15.7	High Wycombe	11.8	14
15	Tamworth	15.3	Tamworth	11.2	15
16	Basingstoke	14.8	Preston	11.1	9
17	Leicester	13.5	Northampton	10.9	11
18	Burnley	13.5	Chester-le-Street	10.5	—
19	Ipswich	13.4	Chesterfield	10.4	—
20	Crawley	13.1	Colchester	10.2	—

Bottom twenty

Rank	City	% 1971	City	% 1981	Rank 1971
20	York	4.0	Bedford	3.4	16
19	Southport	3.6	Teesside	3.4	5
18	Brighton	3.5	Harlow and Bishop's Stortford	3.3	—
17	Norwich	3.5	Hartlepool	3.3	7
16	Bedford	3.4	Plymouth	3.1	—
15	Cardiff	3.4	Brighton	3.0	18
14	SE Fife	3.4	Mansfield	2.9	9
13	Eastbourne	3.3	Norwich	2.9	17
12	Dundee	3.1	Cardiff	2.9	15
11	Carlisle	3.0	Maidstone	2.7	—
10	Falkirk	2.6	Exeter	2.3	4
9	Mansfield	2.2	York	2.2	20
8	Aberdeen	2.1	Greenock	2.2	—
7	Hartlepool	2.0	Aberdeen	2.2	8
6	Edinburgh	1.9	Southport	2.1	19
5	Teesside	1.6	Scunthorpe	1.6	1
4	Exeter	1.4	Dundee	1.6	13
3	Chester	1.3	Chester	1.2	3
2	Torbay	0.7	Torbay	1.1	2
1	Scunthorpe	0.4	Edinburgh	0.9	6

Great Britain % 8.6 (1971) 6.7 (1981)

Table 3.A5. *Employment in the Top and Bottom Twenty Cities in Industry Group E as a Percentage of Total Employment in 1971 and 1981*

Top twenty

Rank	% 1971	City	% 1981	Rank 1971
1	23.9	Chelmsford	19.5	2
2	22.1	Hertford	17.2	3
3	21.6	St Albans	15.4	6
4	18.2	Hamilton and E. Kilbride	13.2	1
5	17.7	Harlow and Bishop's Stortford	12.8	7
6	17.4	Derby	11.0	—
7	17.3	Coventry and Nuneaton	10.6	17
8	16.5	Basildon	10.3	—
9	14.7	Airdrie and Coatbridge	10.3	4
10	13.4	Guildford	9.7	—
11	11.9	Cheltenham	9.6	—
12	11.3	Hartlepool	9.6	14
13	11.2	Basingstoke	9.5	13
14	11.2	Crawley	9.5	—
15	9.3	Bristol	9.4	10
16	9.0	Brighton	9.3	—
17	9.0	Hemel Hempstead	8.9	—
18	8.9	Swindon	8.0	11
19	8.9	SE Fife	8.8	8
20	8.8	Aldershot	8.6	—

Great Britain % 4.9 (1971) 4.4 (1981)

Bottom twenty

Rank	City	% 1971	City	% 1981	Rank 1971
20	Falkirk	1.1	Peterborough	1.2	—
19	Sheffield	1.0	Cardiff	1.2	—
18	Harrogate	1.0	Morecambe	1.2	8
17	Maidstone	0.9	Teesside	1.2	—
16	Mansfield	0.8	Aberdeen	1.2	12
15	Chester-le-Street	0.8	Maidstone	1.1	17
14	Telford	0.7	Chester	1.0	—
13	Swansea	0.5	Swansea	0.9	13
12	Aberdeen	0.5	Barnsley	0.9	—
11	Kettering	0.4	Warrington	0.8	6
10	Tamworth	0.4	Scunthorpe	0.8	1
9	Burton upon Trent	0.4	Wakefield and Dewsbury	0.7	—
8	Morecambe	0.4	Falkirk	0.7	20
7	Newport	0.3	Mansfield	0.5	16
6	Warrington	0.3	Doncaster	0.5	4
5	Darlington	0.3	Blackpool	0.4	—
4	Doncaster	0.2	Carlisle	0.3	—
3	Cannock	0.0	Grimsby	0.2	—
2	Dearne Valley	0.0	Darlington	0.2	5
1	Scunthorpe	0.0	Dearne Valley	0.1	2

Table 3.A6. *Employment in the Top and Bottom Twenty Cities in Industry Group F as a Percentage of Total Employment in 1971 and 1981*

Top twenty

Rank		% 1971		% 1981	Rank 1971
1	Aldershot	19.1	Crawley	19.6	2
2	Crawley	13.9	Aldershot	18.8	1
3	London	13.7	London	17.3	3
4	Harrogate	9.7	Reading	13.6	7
5	Bedford	9.5	Chester	12.3	—
6	Edinburgh	9.2	Southend	12.1	17
7	Reading	8.8	Edinburgh	12.0	6
8	Cheltenham	8.3	Brighton	11.8	11
9	Blackpool	8.2	Norwich	11.7	10
10	Norwich	7.9	Cheltenham	11.5	8
11	Brighton	7.7	Guildford	11.4	14
12	Harlow and Bishop's Stortford	7.6	Slough	10.5	—
13	Cambridge	7.6	Bristol	10.3	—
14	Guildford	7.5	Warrington	10.2	16
15	Bournemouth	7.5	High Wycombe	10.1	—
16	Warrington	7.3	Bournemouth	10.1	15
17	Southend	7.1	Ipswich	10.1	—
18	Northampton	6.5	Northampton	9.8	18
19	Exeter	6.1	Luton	9.7	—
20	Hertford	6.0	Blackpool	9.7	9

Bottom twenty

Rank		% 1971		% 1981	Rank 1971
20	Potteries	2.3	Grimsby	4.1	—
19	Doncaster	2.3	Hamilton and E. Kilbride	4.1	—
18	Sunderland	2.3	The Potteries	3.9	20
17	Chesterfield	2.2	Scunthorpe	3.9	13
16	Swindon	2.1	Derby	3.8	—
15	Wakefield and Dewsbury	2.1	Dearne Valley	3.8	1
14	Tamworth	2.0	Blackburn	3.5	—
13	Scunthorpe	2.0	Chesterfield	3.4	17
12	Greenock	2.0	Morecambe	3.3	—
11	Thames Estuary	1.9	SE Fife	3.2	—
10	Chester-le-Street	1.9	Wakefield and Dewsbury	3.1	15
9	Hartlepool	1.8	Falkirk	3.1	8
8	Falkirk	1.7	Burton upon Trent	3.0	—
7	St Helens	1.6	Doncaster	3.0	19
6	Barnsley	1.6	Greenock	3.0	12
5	Motherwell and Wishaw	1.5	Motherwell and Wishaw	2.9	5
4	Airdrie and Coatbridge	1.3	Mansfield	2.8	3
3	Mansfield	1.2	Hartlepool	2.5	9
2	Cannock	1.1	Barnsley	2.4	6
1	Dearne Valley	0.9	Cannock	1.9	2

Great Britain %   6.1 (1971)   8.5 (1981)

Table 3.A7. *Employment in the Top and Bottom Twenty Cities in Industry Group G as a Percentage of Total Employment in 1971 and 1981*

Top twenty

Rank		% 1971		% 1981	Rank 1971
1	Torbay	49.0	Torbay	49.5	1
2	Eastbourne	39.7	Southport	39.9	3
3	Southport	38.0	Eastbourne	38.2	2
4	Bournemouth	33.9	Blackpool	38.1	5
5	Blackpool	32.8	Sunderland	38.0	20
6	Brighton	31.4	Bournemouth	36.3	4
7	Hastings	31.1	Hastings	33.9	7
8	Thanet	29.0	Chester	33.8	9
9	Chester	28.8	Hamilton and E. Kilbride	33.1	—
10	Southend	28.4	Harrogate	32.9	11
11	Harrogate	28.4	Basingstoke	32.0	—
12	Exeter	28.2	Swindon	31.3	—
13	Aberdeen	28.2	Worcester	30.7	17
14	Plymouth	27.6	Brighton	30.2	6
15	Grimsby	27.5	Aberdeen	30.2	13
16	Cheltenham	27.0	Hull	30.1	—
17	Worcester	26.9	Southend	30.0	10
18	Edinburgh	26.8	Grimsby	29.9	15
19	Morecambe	26.7	Colchester	29.8	—
20	Sunderland	26.0	Tamworth	29.8	—

Bottom twenty

Rank		% 1971		% 1981	Rank 1971
20	Blackburn	17.1	Burton upon Trent	23.5	—
19	Hartlepool	16.6	Telford	24.4	7
18	Luton	16.4	Basildon	23.0	6
17	Harlow and Bishop's Stortford	16.4	Burnley	23.0	—
16	The Potteries	16.4	Huddersfield	22.5	9
15	St Helens	16.4	Birmingham	22.2	—
14	Motherwell and Wishaw	16.3	Greenock	22.2	—
13	Derby	16.2	The Potteries	22.1	16
12	Scunthorpe	16.1	Hartlepool	22.1	19
11	Milton Keynes	16.0	Coventry and Nuneaton	21.4	10
10	Coventry and Nuneaton	15.9	Doncaster	21.0	—
9	Huddersfield	15.9	Barnsley	20.8	—
8	Kettering	15.4	Blackburn	20.5	20
7	Telford	15.4	Motherwell and Wishaw	20.0	14
6	Basildon	15.3	Cannock	19.7	5
5	Cannock	15.0	Halifax	18.1	3
4	Chesterfield	14.7	Chesterfield	18.0	4
3	Halifax	13.9	Mansfield	17.8	2
2	Mansfield	13.5	Derby	17.5	13
1	Dearne Valley	10.8	Dearne Valley	17.2	1

Great Britain %   22.4 (1971)   26.6 (1981)

Table 3.A8. *Employment in the Top and Bottom Twenty Cities in Industry Group H as a Percentage of Total Employment in 1971 and 1981*

Top twenty

Rank		% 1971		% 1981	Rank 1971
1	Oxford	32.8	Exeter	33.7	2
2	Exeter	30.2	Cambridge	33.4	3
3	Cambridge	30.0	Oxford	32.1	1
4	Morecambe	27.6	Morecambe	29.5	4
5	Chelmsford	25.9	Eastbourne	28.3	9
6	Maidstone	25.8	Dundee	28.3	—
7	Bath	25.6	Maidstone	27.1	6
8	York	25.1	Edinburgh	25.7	16
9	Eastbourne	24.8	Wakefield and Dewsbury	25.3	11
10	Cardiff	24.6	Milton Keynes	25.2	—
11	Wakefield and Dewsbury	24.6	Bedford	25.2	—
12	Chester	24.1	Airdrie and Coatbridge	25.0	—
13	Reading	23.8	Hastings	24.8	15
14	Southport	23.7	Southport	24.7	14
15	Hastings	23.5	Harrogate	24.7	18
16	Edinburgh	23.5	Derby	24.6	—
17	Colchester	23.3	Lincoln	24.5	—
18	Harrogate	23.1	Bath	24.2	7
19	Thanet	22.6	Cardiff	24.1	10
20	Hamilton and E. Kilbride	22.5	Swansea	23.9	—

Great Britain %   17.9 (1971)   20.4 (1981)

Bottom twenty

Rank	% 1971			% 1981	Rank 1971
20	Peterborough	14.9	High Wycombe	17.0	—
19	Grimsby	14.6	Southampton	16.8	10
18	Kettering	14.4	Telford	16.5	12
17	Leicester	14.3	Burnley	16.5	—
16	Tamworth	14.3	Bournemouth	16.4	—
15	Basildon	14.2	Cheltenham	16.4	—
14	Slough	14.0	Chatham	16.1	15
13	Birmingham	13.7	Basildon	15.9	—
12	Burnley	13.6	The Potteries	15.8	—
11	Chester-le-Street	13.5	Scunthorpe	15.8	2
10	Telford	13.3	Luton	15.7	1
9	Aldershot	13.2	Burton upon Trent	15.6	—
8	Mansfield	13.2	Aldershot	15.2	9
7	Airdrie and Coatbridge	13.2	Slough	15.2	14
6	Coventry and Nuneaton	13.1	Peterborough	15.2	20
5	Cannock	12.9	Falkirk	14.9	—
4	Widnes–Runcorn	12.9	Mansfield	14.8	8
3	St Helens	12.8	Cannock	14.4	5
2	Scunthorpe	12.7	Chester-le-Street	13.7	11
1	Luton	11.9	Widnes–Runcorn	13.6	4

Table 3.A9. *Employment in the Top and Bottom Twenty Cities in Industry Group J as a Percentage of Total Employment in 1971 and 1981*

Top twenty

Rank	City	% 1971	City	% 1981	Rank 1971
1	Bath	22.8	Bath	22.0	1
2	Burton upon Trent	20.2	Burton upon Trent	19.7	2
3	Chatham	17.6	Aldershot	17.4	—
4	Cardiff	14.6	Colchester	15.6	7
5	Swansea	14.6	Edinburgh	14.5	—
6	Edinburgh	14.1	London	14.5	5
7	Harrogate	13.7	Cardiff	14.3	13
8	Aldershot	13.6	Portsmouth	14.0	3
9	London	13.4	Hastings	13.5	6
10	Cheltenham	13.3	Carlisle	13.3	—
11	Southend	12.9	Hemel Hempstead	13.2	16
12	Chester-le-Street	12.7	Basingstoke	12.9	—
13	Gloucester	12.7	Harrogate	12.8	17
14	Hastings	12.5	Plymouth	12.5	9
15	Tyneside	12.3	Reading	12.4	—
16	Brighton	11.8	Southend	12.2	—
17	Reading	11.8	Gloucester	12.1	15
18	Hamilton and E. Kilbride	11.8	Chester	12.0	—
19	Warrington	11.6	Telford	11.9	—
20	Chester	11.6	Norwich	11.8	18

Bottom twenty

Rank	City	% 1971	City	% 1981	Rank 1971
20	The Potteries	5.1	Huddersfield	5.8	17
19	Doncaster	5.1	The Potteries	5.6	20
18	Grimsby	5.0	Burnley	5.6	—
17	Huddersfield	4.8	Grimsby	5.6	18
16	Hartlepool	4.7	Falkirk	5.6	12
15	SE Fife	4.5	Basildon	5.5	—
14	Widnes–Runcorn	4.3	Chesterfield	5.5	—
13	Tamworth	4.3	St Helens	5.1	—
12	Falkirk	4.1	Doncaster	5.1	19
11	Chester-le-Street	4.0	Coventry and Nuneaton	4.9	9
10	Hertford	4.0	Airdrie and Coatbridge	4.8	—
9	Coventry and Nuneaton	3.7	Motherwell and Wishaw	4.6	3
8	Barnsley	3.6	Hertford	4.4	10
7	Mansfield	3.5	Scunthorpe	4.4	1
6	Kettering	3.5	Mansfield	4.3	7
5	Airdrie and Coatbridge	3.4	SE Fife	4.1	15
4	Dearne Valley	3.4	Cannock	3.8	2
3	Motherwell and Wishaw	3.3	Barnsley	3.1	8
2	Cannock	3.3	Tamworth	3.0	13
—	Scunthorpe		Dearne Valley	2.3	4

Great Britain %   9.0 (1971)   9.1 (1981)

4

The Beneficiaries of Employment Growth: An Analysis of the Experience of Disadvantaged Groups in Expanding Labour Markets

Nick Buck and Ian Gordon

The focus of the Economic and Social Research Council (ESRC) Inner Cities Research Programme on economic decline in large urban areas, and associated labour-market disadvantage, has meant that the studies are heavily biased towards areas of employment decline, and in fact no areas of significant growth are included. However, some of the research within the programme suggests that the recent experience of employment decline within the study areas may not be the main cause of the disproportionate incidence of unemployment among inner city residents. Rather, it would appear that inner city residents experience continuing disadvantage in the allocation of employment largely as a reflection of personal characteristics—of age, race, marital status, social class, skill, or employment experience—which affect evaluations of their employability. If this, more than employment decline, is the source of chronic inner city unemployment, it raises questions about the potential effectiveness both of policies seeking to relieve unemployment through local employment creation and of policies encouraging the movement of disadvantaged groups away from the areas of decline. In order to determine whether such policies could be either necessary or sufficient conditions for the reduction of unemployment among vulnerable inner city residents, comparative research on areas of employment growth and decline is necessary. It could be, for instance, that policies to improve the relative position of the disadvantaged in the labour market depend for their success on a context of growth, or that *prolonged* employment decline in a local labour market leads to an accumulation of disadvantaging characteristics in the local population.

In order to address this set of questions we need to look at how people with characteristics similar to the characterisitics of those who suffer disadvantage in inner city areas fare where new jobs are being created—examing how well they can compete, both with others in the area and with those outside who might move in, and whether individuals with comparable backgrounds have a greater chance of achieving stable jobs or training in growth areas.

Many inner city residents appear to face a double set of disadvantages in the labour market. As individuals they may possess characteristics—of 'skill';

race, etc.—which worsen their chances in competition for jobs, particularly stable jobs, within the urban labour market. At the same time they find themselves in a spatial labour market (or sub-market) in which employment is declining more rapidly than elsewhere, and where—partly because of unequal constraints on mobility in the housing system—the balance between labour demand and supply may be worsening, particularly for unqualified workers. Of these two sets of factors, the London study (Buck *et al.* 1986) suggests that the former (unequal competition within the market) is the more important source of disadvantage, while the Bristol study (Boddy *et al.* 1986) indicates that great inequalities in employment opportunities may persist in more 'successful' local labour markets. Nevertheless, we should expect variations in employment growth rates to have rather more effect on local unemployment at the bottom end of the labour market. Certainly local unemployment rates, at least for males, tend to vary much more widely among the unskilled, although it is not clear to what extent this reflects their lesser mobility or a 'bumping-down' process in which the ultimate effects of employment fluctuations in other occupational groups get transmitted to those in the weakest competitive position.

The question of how far the benefits of employment growth in an area are experienced by the less advantaged residents of that area is of significance for two reasons. First, it bears on the policy question of how far job creation policies in the inner city can reduce unemployment among local residents, and inequality in the distribution of unemployment within the city. Secondly, it bears on the policy question of how far active policies of dispersal of the disadvantaged to areas with more favourable growth performance would improve their employment chances as well as improving the supply–demand balance for those remaining in the inner areas.

For both of these policy questions it is important first of all to establish what sort of employment is involved and how accessible this is to disadvantaged groups. As far as the dispersal policy is concerned, this should be essentially a question of fact—namely, the extent to which employment opportunities for groups without qualifications etc. have actually been created in areas of employment growth outside the cities. As far as job creation policies for the inner city are concerned, the issue is rather more one of identifying the range of types of employment where growth would significantly affect unemployment among local residents.

'Disadvantage' in the labour market is a loose and rather unsatisfactory concept which does, however, have the virtue of implying that there is no precise set of personal characteristics which determine *ex ante* a disadvantaged status in the labour market. Disadvantage in access to employment opportunities and exposure to unemployment is the result of a cumulative set of processes. At a conceptual level is it therefore necessary to distinguish between *background* influences, in terms of which a large proportion of workers are potentially disadvantaged, and the *immediate* factors in terms of which particular minorities of workers are actually disadvantaged. In the context of a specific labour market several of the same characteristics—for example, race—may contribute both indirectly and directly to an individual's chances of, for example, becoming unemployed, and so no simple division can be made between those who are 'potentially' and 'actually' disadvantaged. However, it is an important fact that actual disadvantage is to a large extent a function of the particular set of jobs that an individual has come to occupy, and that this status is the outcome of chance processes, conditioned by a set of background circumstances. Since this chapter is concerned with the conditioning effect of local labour-market circumstances on the probabilities, firstly, that 'potentially disadvantaged' individuals are actually 'disadvantaged' and, secondly, that 'disadvantaged' individuals are unemployed, locational characteristics will not themselves be used as indicators of either actual or potential disadvantage. For rather different reasons, age will also be excluded from the set of disadvantaging characteristics. Both the youngest and the oldest working age groups can experience disadvantage in gaining employment, particularly in circumstances of low demand for labour, but the majority of young workers (in the 18–30 age range) face an above-average chance of unemployment more as a reflection of their relative mobility in the labour market than owing to structural disadvantage. Nevertheless, it will be appropriate to disaggregate some of our analyses to see how other potentially disadvantaging characteristics and locational factors interact to affect current disadvantage and unemployment among new entrants to the labour market and the over-55s.

The basic set of relationships with which we are concerned is summarized in the form of a path diagram in Figure 4.1. Within the context of this study it has not been feasible to model this complete set of relationships, partly because of the lack of suitable longitudinal data. However, the diagram does serve to indicate the set of relationships which we shall have to take into account. In particular, it suggests that area influences (including the effect of employment growth/decline) may enter at three different stages, affecting the possession/non-possession of qualifications, stability/instability of employment, and the risk of unemployment. The London study in this programme (Buck *et al.* 1986) found evidence that (when other factors were held constant) Inner London residents were disadvantaged at all three stages. This may not in each case have been the consequence of employment decline *per se*—and, in the case of qualifications levels, current conditions (and even the current area of residence) may be less relevant than those obtaining at some time in the past.

The remainder of this chapter is in five sections. The first of these distinguishes a number of sets of areas in which some employment growth has been occurring in the recent past and examines the form which that growth has taken. The following section is more theoretical in character, analysing the causal relations which are to be expected between local employment

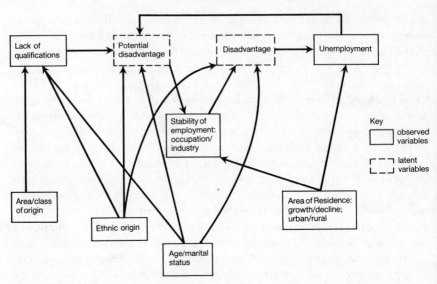

Fig. 4.1. Path Diagram summarizing the Relationships involved in Disadvantage in the Labour Market.

growth and the incidence of unemployment, particularly among disadvantaged groups. In the light of these expectations the third section examines the actual labour-market experience of these groups in the various growth areas. The fourth section focuses more specifically on the relationship between individuals' characteristics and the risk of unemployment, and on possible differences in that relationship between growing and declining areas. This contrast is followed up in the fifth section which explores the extent to which differing labour-force characteristics may themselves have been generated by sustained experience of growth or decline.

Areas of Employment Growth

Over the period since the 1960s employment growth in Britain as a whole has been sluggish at best, and net losses have been experienced since the early 1970s. Employment decline accelerated sharply in the recession after 1979, and total employment had fallen by nearly 9 per cent in the period up to 1984. However, underlying this historical pattern there have been two dominant spatial components of much more persistent importance, though with some variation in impact in different periods. These have led to very substantial variation in the rate of employment change in different areas.

The first component has been a process of counter-urbanization, involving both a decentralization of employment within city regions and a tendency for the smaller city regions to grow faster than the larger. This process—documented in Hall *et al.* (1973) and Spence *et al.* (1982)—has affected both population and employment, with some interaction between changes in the two. Fothergill and Gudgin (1982) suggest that a key factor has been the redistribution of manufacturing investment towards areas which can more readily accommodate the sector's increasing demand for space. Spatial cost differences for both land and labour have also been important, affecting many service activities as well as manufacturing. Both these factors of production tend to be cheaper in less urban areas because of weaker demand in the past, and population shifts themselves have provided an elastic source of labour supply in these areas. Household movements have also served to induce a counterurbanization of employment in consumer services. This process has not, however, operated uniformly across time or space. It seems to have been at its height in the second half of the 1960s and first half of the 1970s, when economic growth was stimulating both industrial investment and the private housing market, and has slowed down substantially during the current recession. For related reasons, the process of counter-urbanization has been rather more evident in the southern half of the country.

The second process has involved a marked shift in the balance both of population and of economic activity from North to South. This has been in part a consequence of the spatial distribution of declining industries, but also reflects biases in the location of new investment and the apparent pattern of residential preferences. The southward shift has been partly offset by the decline of London, and was reversed for a period in the early 1970s, when this decline was at its height and regional policy was having its greatest effects. However, since the mid 1970s and particularly in the current recession, the bias in growth against the North has become much more marked, and it is almost exclusively in the three southern regions—the South East, East Anglia, and the South West—that areas of employment growth are to be found.

The combined impact of these two processes can be seen in employment changes between 1976 and 1981. Over this period, total employment in Great Britain declined by almost three-quarters of a million. This net change was more than accounted for by a decline of over 900 000 jobs in the conurbations. Employment in the non-conurbation areas actually increased, but this was entirely attributable to changes in the southern half of the country. Even at the regional level there have been substantial variations in the rate of employment change over recent years. Table 4.1 ranks the regions by employment change during the period 1971–81, confirming the distinctly favourable performance of the southern regions, except for London. As it shows, male employment declined in every region except East Anglia and the Rest of the South East (excluding London); female employment, on the other hand, grew in every region except Greater London.

In order to explore the labour-market conditions associated with employment growth, and compare them with those obtaining in areas of decline, a set of growth areas was selected on the basis of employment change between

Table 4.1. *Regional Employment Change in Great Britain, 1971–1981*

Region	Total (%)	Male (%)	Female (%)
East Anglia	+12.2	+3.4	+25.4
Rest of the South East	+11.2	+2.5	+24.8
South West	+8.2	−1.0	+23.4
East Midlands	+4.0	−3.8	+17.4
Scotland	−0.6	−8.0	+10.9
Great Britain	−1.5	−8.9	+10.5
Wales	−2.6	−13.1	+17.1
Yorkshire and Humberside	−3.9	−11.1	+8.4
North	−7.3	−15.4	+7.4
North West	−7.6	−14.8	+3.7
West Midlands	−7.9	−14.6	+3.5
Greater London	−9.5	−14.2	−2.1

Source: Censuses of Employment 1971 and 1981.

the 1971 and 1981 Censuses of Population. In general the criterion for inclusion was that a district or group of districts approximating a labour-market area should have had an increase in male employment over the decade—a relatively restrictive criterion ensuring substantial overall growth. With one significant exception, the search for such areas was restricted to the southern and eastern regions which had the most favourable employment performance. Since the data to be used for most of the analysis were derived from the 1981 European Economic Community Labour Force Survey (LFS), a further criterion for inclusion was that the labour-market area should include districts sampled by the LFS. A number of areas with growing employment had to be excluded.

In order to explore conditions in different types of area, the growth areas were assigned to six groups:

(I)   *Expanding city regions*—that is, growing areas which have at their core one or more relatively large cities. Both of the cases falling in this category were areas for which considerable population and employment growth was planned in the 1960s. In practice, neither South Hampshire nor Severnside quite met the criterion of increasing male employment, as growth on the periphery has not been sufficient to balance losses in the urban core. The aggregate employment change in both areas has, however, been much more favourable than that in other large cities in the country, and it appeared to be worth studying how the more disadvantaged have fared in these labour markets.

(II) *Large southern New Towns.* The major cases with growth in the 1970s were Milton Keynes, Northampton, and Peterborough. The distinctive feature of this group is the very high rate of growth of *both* employment and population.

(III) *Growing northern and midland New Towns.* These are the exceptions referred to above. Their significance is that, while they have been growing, their surrounding regions have been contracting quite sharply. The cases included are Warrington and Telford.

(IV) *Other growing urban centres and their hinterlands.* These are areas with a relatively large urban core (having population of 100 000 or so) and growing employment over the labour-market area as a whole. In addition to their differences in planning status from the New Towns (notably the absence of planned population movements), these also differ from the two previous groups in that the urban core itself is not necessarily growing, and that there may have been significant decentralization in the labour market. The major cases here are Plymouth, Reading, Cambridge, Ipswich, and Norwich.

(V) *Rural growth areas.* These are distinctive in not containing substantial urban areas. The major representation of these areas is in East Anglia and parts of the South West—that is, Somerset, Dorset, and South Devon.

(VI) *The metropolitan fringe.* There is employment growth in some areas on the edge of London, including some outer London boroughs. In most cases these clearly do not constitute independent labour markets, but the situation of the disadvantaged in these areas is of interest, since their growth reflects the major form of unplanned decentralization from London.

The areas included in each group are shown in Figure 4.2, and the employment changes in each are shown in Table 4.2. Because the areas were selected on the basis of employment growth, their rates of change cannot be taken as evidence for or against counter-urbanization hypotheses, though in fact growth was much lower in the city regions than in the other groups, and most of the areas are not attached to larger cities.

We have already noted the difference in the rate of employment change

Table 4.2. *Employment Change in the Selected Growth Areas, 1971–1981*

Area	Total (%)	Male (%)	Female (%)
City regions	+4.5	−1.8	+16.1
Southern New Towns	+29.8	+22.4	+42.8
Northern New Towns	+8.0	+1.5	+21.3
Urban areas	+12.5	+6.5	+23.5
Rural areas	+10.7	+4.1	+22.7
Metropolitan fringe	+7.4	+3.4	+13.4
Inner London	−15.1	−16.0	−13.3
Greater London	−11.7	−13.7	−8.6
Other conurbations	−10.4	−15.8	−1.3

Source: Censuses of Population 1971 and 1981.

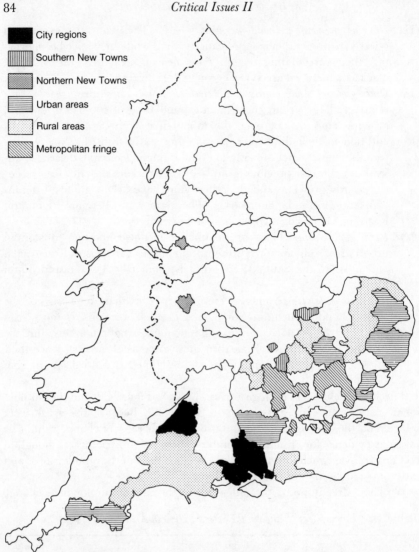

Fig. 4.2. The Selected Areas of Employment Growth.

between men and women, and it is again apparent for the growth areas from
Table 4.2. While employment growth for women was relatively general during the 1970s, areas where male employment grew were much fewer. There
were also considerable variations in the rate of employment change by industry and by occupation. This section concentrates on the latter, since an occupational analysis can be related more closely to the employment
opportunities of those experiencing disadvantage in the labour market.
These occupational changes are in some cases primarily a consequence of
industrial employment changes, whereas others reflect shifts in occupational

composition within a range of industries. Of the two most important changes at the national level, the decline in employment in manual occupations has been largely a consequence of contractions in manufacturing, mining, construction, and transport industries, where manual workers are in the majority. On the other hand, the growth of administrative, professional, and technical (APT) jobs, has mostly followed from changes in occupational mix within industries. At the national level, APT employment grew by 28 per cent between 1971 and 1981, clerical employment by 3 per cent and employment of sales and service workers by 4 per cent, while manual employment declined by 19 per cent. The decline in manual employment and the growth in service employment have proceeded in parallel for men and women. Growth in APT employment, on the other hand, has been substantially faster for females, while there has been a marked substitution of women for men in clerical employment. Gross changes in employment for occupational groups have thus been much greater than the net change in overall employment, and these have had important consequences for groups of workers whose access to non-manual jobs is limited. Not only has the pool of manual jobs been substantially reduced, but many of the service jobs which have become available for the unqualified have offered inferior employment conditions in terms of stability of tenure and training opportunities. In the London study, for example, we noted that such workers had become increasingly dependent on the public sector for more stable jobs as manufacturing employment had shrunk, but that this source of employment had also started to decline (Buck *et al.* 1986, 172–3).

One of the key issues at the subregional level, then, is whether the pattern of occupational shifts has been the same or different in the growth areas, and particularly whether there has been any significant growth in manual employment. In fact, for areas where it is possible to measure change[1] it appears that national patterns of relative change of occupation have generally been reproduced, but with more favourable changes in each occupation. In all the cases shown in Table 4.3, manual employment change was below total change, and for men the margin is reasonably consistent at around 10 per cent, with apparently more favourable manual performance in Huntingdonshire and West Sussex (the latter may be unevenly affected by a boundary change involving Gatwick Airport) and worse relative performance in the Berkshire, Buckinghamshire, and Oxfordshire group and the southern New Towns. The New Towns were, however, notable as being the only areas where the overall level of growth was sufficient to yield a substantial increase in male manual employment. The pattern of change for the other occupation groups in all areas tended to reflect the national pattern outlined above. In every case the most rapidly expanding group of male workers was the APT group, and this was true for female workers in all but one case. Service occupations were generally the second most rapidly

Table 4.3. *Percentage Change in Employment by Occupation in Growth Areas,*
1971–1981

	Total	APT	Clerical	Service	Manual
Males					
Berks., Bucks. and Oxfordshire	+4.4	+43.6	−4.2	+14.7	−9.7
Cambridgeshire[a]	+0.7	+25.1	−8.8	+15.9	−10.0
Dorset[a]	+6.2	+43.2	+15.9	+11.9	−7.3
Hampshire[b]	−0.3	+32.0	−5.7	+7.7	−11.5
Huntingdonshire[a]	+10.1	+50.4	+2.1	+18.3	+1.7
Norfolk	+2.6	+46.7	+2.2	+10.2	−8.0
Suffolk	+3.6	+42.8	+1.9	+16.6	−6.7
West Sussex[a]	+8.5	+39.8	−1.7	+9.3	+1.1
Southern New Towns[c]	+22.4	+74.5	+13.4	+33.5	+7.0
Northern New Towns[d]	+1.5	+60.3	−16.2	+14.5	−10.9
Females					
Berks., Bucks. and Oxfordshire	+21.6	+62.6	+34.4	+10.3	−11.0
Cambridgeshire[a]	+17.6	+47.9	+27.8	+3.6	−0.4
Dorset[a]	+25.6	+51.2	+57.4	+12.8	+8.2
Hampshire[b]	+21.0	+61.8	+41.1	+8.5	−13.6
Huntingdonshire[a]	+41.5	+80.4	+52.3	+27.9	+21.9
Norfolk	+18.5	+68.5	+37.7	+14.4	−12.6
Suffolk	+22.9	+60.4	+35.2	+16.5	−3.0
West Sussex[a]	+23.0	+68.4	+34.3	+15.8	−10.2
Southern New Towns[c]	+42.8	+101.0	+62.4	+40.6	−5.1
Northern New Towns[d]	+21.3	+87.3	+39.7	+23.0	−23.7

[a] 1971 Definition of county.
[b] 1971 Definition of county less Bournemouth.
[c] Milton Keynes, Northampton, and Peterborough.
[d] Telford and Warrington.

Source: Censuses of Population 1971 and 1981.

expanding group for men, but clerical employment grew more rapidly than
service jobs in the case of women.

So far this section has considered the difference between areas simply in
terms of variations in the rate of employment growth. However, while, other
things being equal, we should expect a lower rate of unemployment in areas
of employment growth, there are important respects in which other things
are not equal. In particular, there are considerable variations in the rate of
population expansion, and hence in the growth of the labour force, which
must also affect the balance of supply and demand in local labour markets.
Owen *et al.*'s (1984) classification of local labour-market areas in Britain dis-
tinguishes three types of areas with male employment growth. Of these, the
cluster with the most rapid growth also had the highest rate of unemployment growth (a growth close to that experienced in areas of rapid employment decline) because of an even more rapid increase in labour supply as a
consequence of inward migration. This cluster consists mainly of New and

Expanded Towns, including the third-generation New Towns of Milton Keynes and Peterborough, included in our Group II, as well as Telford from Group III. The most favourable labour-market conditions were found in the cluster experiencing slow growth, which included most of the metropolitan fringe, some of the larger southern cities, and large areas of rural East Anglia and the South West. In these areas the rate of labour supply growth was slow enough to ensure the lowest rate of unemployment growth of any of their seven clusters. A third cluster, with a broadly similar level of employment growth, but a higher rate of natural increase, experienced a rather higher rate of unemployment growth. This cluster comprised most of the older south-eastern New Towns (where planned in-migration had ceased), and much of the postwar growth area to the north and west of London stretching from Reading to Bedford.

The contrasting experience of these three sets of growth areas indicates that the link between employment growth and low or falling rates of unemployment cannot be taken for granted. Before looking in more detail at the experience of unemployment in the various growth areas, we need therefore to look more closely at the causal links between employment change, labour supply changes, and unemployment, and their implications for particular subgroups of workers including the disadvantaged.

The Relationship between Employment Change and Unemployment

Employment growth in local labour markets has both direct and indirect implications for local unemployment, the strength of which depends largely on whether or not growth is also occurring in other areas, particularly those within the same region. The immediate effect of employment changes in any area is partly absorbed by shifts in the balance of migration and/or of commuting flows, as well as by individuals moving in or out of the labour force—with a residual impact on local unemployment varying in scale according to the degree of labour-market closure of the area. In subsequent periods the change in local unemployment relative to that in other areas will give rise to further adjustments in the balance of labour migration, thus progressively reducing the impact of the original change. Once-for-all changes in employment levels within an area should thus have only temporary effects on local unemployment; correspondingly, sustained differentials in rates of employment change are expected to induce, not a continuous divergence in unemployment, but a stable set of differentials in unemployment rates. Hence, if there were no variation in the incidence of structural or frictional unemployment, and no other sources of variation in the growth of labour supply, local unemployment rates would vary inversely with the rate of employment change in an area. However, the extent of this variation would depend on levels of mobility in the labour force. At times, such as the current recession, when geographical mobility is inhibited, the spread of unemployment rates between areas would tend to widen (Gordon 1985). Similarly, among groups of workers, such as the less-skilled, whose mobility is less, or less responsive to local employment growth, there should be greater disparities in unemployment rates as between areas of growth and of decline. Groups that are less mobile, in terms of both labour migration and commuting, ought thus to be more affected in terms of their chances of unemployment by rates of local employment change.

This hypothesis can be expressed more clearly in terms of a simple model of equilibrium unemployment rates.[2] Changes in unemployment within a particular occupational group k in a particular area i can be represented as the outcome of changes in the level of employment, natural change in labour supply, net labour migration, changes in the net balance of commuting, and net shifts by local workers out of other occupational groups:[3]

$$\Delta UE_{ik} \equiv NI_{ik} + NM_{ik} + \Delta NC_{ik} + NO_{ik} - \Delta E_{ik} \qquad (4.1)$$

Where UE is the number of unemployed, NI the natural increase in economically active (within occupation group), E the number in employment, NM the net labour migration from other areas, NC net commuting into the area, and NO the net movement into occupational group by economically active residents.

This identity may be rewritten in terms of rates with variables being expressed as differences from those in a corresponding national equation:[4]

$$\Delta(U_{ik} - U_{nk}) \equiv s_{ik} - r_{ik} + m_{ik} + \Delta c_{ik} + o_{ik} \qquad (4.2)$$

Where U is the unemployment rate, s the regional deviation from the national rate of natural change in labour supply, r the regional deviation from the national rate of employment change, m the rate of net inter-regional labour migration, c net inward commuting, o the regional deviation from the national rate of net inward movement to the occupation, the subscript n identifies a national value, and all rates are expressed as proportions of the economically active in the occupational group.

Labour migration, net commuting, and occupational mobility are all liable to be affected by the level of unemployment and/or employment changes: in the first two cases what is relevant are differences in conditions between the area and its (migration or commuting) hinterland; in the last case it would be differences between occupational groups within the area—for example,

$$m_{ik} = \alpha_k(U_{ik} - U_{mk}) + \beta_k(r_{ik} - r_{mk}) + \gamma(A_{ik} - A_{mk}) + \varkappa_k(X_i - X_m) \quad (4.3)$$

$$\Delta c_{ik} = C_{ik}(r_{ik} - r_{ck}) \qquad (4.4)$$

$$o_{ik} = \lambda_k(r_{ik} - r_{io}) + \mu_k(U_{ik} - U_{io}) \qquad (4.5)$$

Where A_{ik} is a summary indicator of the competitive advantage of area i's workers for jobs in occupation k; X is a summary indicator of an area's attractiveness to potential migrants in terms other than employment availability; C

is the mean of in and out commuting as a proportion of economically active; the subscripts m, c, and o denote average values for other areas (for m and c) or occupations (for o), weighted by the volume of flows (labour migration, commuting or occupational movement) between each and occupation k in area i; α, β, γ, $\varkappa$, λ, and μ are coefficients.

In practice, as we note elsewhere, levels of net migration depend also on the accessibility of areas (see Burridge and Gordon 1981), commuting changes also reflect the balance of local residential movements (Gordon and Molho 1985), and occupational changes are influenced by factors such as relative earnings (Gordon and Molho, forthcoming). These elaborations are not important in the present context, however.

An expression for the equilibrium unemployment rate may be derived by substituting the behavioural equations (4.3), (4.4), and (4.5) into the identity (4.2) and setting the *change* in relative unemployment equal to zero. By solving this equation we can obtain an expression for the unemployment rate in particular areas and occupational groups which would obtain if induced migration, commuting, and occupational changes occurred with no significant lag—or if differences in rates of employment growth and natural change had persisted for long enough to allow these flows to adjust fully. In other circumstances, the determinants of actual unemployment rates would involve a weighted average of past changes in employment and labour supply, and of unemployment rates elsewhere, rather than simply the current values.

Initially we shall assume that there is no induced mobility between occupations within an area. In this case, the equilibrium unemployment rate is given by

$$U_{ik} = U_{mk} + \frac{r_{ik} - s_{ik}}{\alpha_k} - \frac{\beta_k}{\alpha_k}(r_{ik} - r_{mk}) - \frac{C_{ik}}{\alpha_k}(r_{ik} - r_{ck})$$

$$- \frac{K_k}{\alpha_k}(X_i - X_m) - \frac{\gamma_k}{\alpha_k}(A_{ik} - A_{mk}) \tag{4.6}$$

For all occupational groups, unemployment in the area is determined by a combination of: the average unemployment rate in its migrational hinterland, differences from both the migrational and commuting hinterlands in rates of employment change and natural change in labour supply, as well as other factors attracting net labour migration and compositional characteristics affecting the competitiveness of the local labour force. Environmentally attractive areas or places such as New Towns where in-migration is being deliberately stimulated will tend to have higher unemployment rates than their rate of employment increase would suggest. The same would be the case for areas of localized growth within declining regions.

The impact of local employment and supply changes must depend substantially on the proportion of workers commuting from outside the area *and* the responsiveness of labour migrants in the particular occupational group to differentials in employment change and unemployment. In relation to commuting, theory suggests that the willingness of workers to commute longer distances should depend largely on their earning capacity, with higher status groups being expected to commute rather further. This pattern generally holds in practice, although in some cases unskilled manual workers may have to commute further than the semi-skilled. In relation to labour migration, mobility levels are highest among professionals and lowest among manual workers (although not particularly the unskilled). The pattern of movement of the low-skilled and of females (both single and married) is less directly related to rates of local employment change than is the case for groups with ·more organized national labour markets. There is thus a clear expectation that the relationship between employment change and unemployment at the local level will be strongest at the bottom end of the labour market, for unskilled, low-paid, or secondary sector jobs.

The effect of dropping the assumption that there is no induced occupational movement when employment grows or declines faster in some occupational groups than in others is to make unemployment in each occupation partly dependent on employment change and unemployment in other occupations. The full specification for the equilibrium unemployment rate equation is then

$$U_{ik} = \frac{\alpha_k}{\alpha_k + \mu_k} U_{mk} + \frac{\mu_k}{\alpha_k + \mu_k} U_{io} + \frac{r_{ik} - s_{ik}}{\alpha_k + \mu_k} - \frac{\beta_k}{\alpha_k + \mu_k}(r_{ik} - r_{mk})$$

$$- \frac{\lambda_k}{\alpha_k + \mu_k}(r_{ik} - r_{io}) - \frac{C_{ik}}{\alpha_k + \mu_k}(r_{ik} - r_{ck})$$

$$- \frac{\varkappa_k}{\alpha_k + \mu_k}(X_i - X_m) - \frac{\gamma_k}{\alpha_k + \mu_k}(A_{ik} - A_{mk}) \quad (4.7)$$

Unemployment in particular labour sub-markets should thus reflect that experienced in neighbouring areas *and* occupation groups, while the impact of employment change in the focal area and occupational group is liable to be weakened by spillovers both into other areas and into other occupational groups. Empirical evidence that occupational, as well as areal, spillovers are important even for very broadly defined groups is provided by an analysis of male unemployment in four social class groupings across areas within the London region, showing that in each case the strongest influence was unemployment in an adjoining class group (Gordon 1981). In growth areas, therefore, we may expect unemployment in all or most occupational groups to be below the national average, particularly if they are surrounded by other growth areas, even if employment change is not particularly favourable in each of the occupational groups.

An implicit assumption in this simple model[5] is that disadvantaging characteristics in the labour market have a constant effect on the probability of unemployment, independent of the general level of unemployment. This is a convenient approximation, but in practice it appears a rather unrealistic assumption. In a tight labour market, employers clearly have less choice as to who they employ, and the gap in unemployment rates between supposedly 'superior' and 'inferior' groups of workers is likely to be less than in a slack labour market. Indeed, in the latter situation, when there may be very large numbers of applicants for each job, coarse filtering criteria relating to age, class, gender, and educational qualifications are likely to play a much more important role in narrowing the field for selection than would be the case when jobs were more readily obtained (see Gordon 1984).

A more plausible assumption is that the effect of personal characteristics on the probability of being unemployed is *proportional* rather than additive: for example, one group may be *twice* as likely to be unemployed as another, rather than always having (say) a 5 per cent greater chance of being out of work. This assumption would be consistent with the evidence[6] that, at constituency level within Greater London, increases in unemployment among residents between the 1971 and 1981 Censuses were essentially proportional to the areal incidence of unemployment in the first year. Similarly, comparison of the logit analyses of individuals' unemployment in 1981 (reported in Tables 4.7 and 4.8 below) with corresponding analyses for Londoners in 1979 suggests that the variation in odds-ratios between groups has neither widened nor narrowed in the face of a general doubling of unemployment.[7]

There are thus two basic reasons for expecting that disadvantaged groups in the labour market may benefit particularly from living in an area in which employment growth is occurring. In the first place this would tend to be the case because they are operating in labour sub-markets in which geographical mobility is more restricted and information about vacancies is less widely diffused. Consequently *local* workers are rather more likely to feel the effects of employment changes than would be the case among higher status groups. In the second place, provided that the balance between unemployment and vacancies is actually more favourable in the growth areas, disadvantaged groups should be in a relatively stronger position to compete for any particular type of job when potential employers have smaller numbers of applicants to choose between.

The main qualification to both of these points is that in all labour markets the extent of mobility is such that much of the effect of employment change— particularly of employment *growth*—will be dispersed outside the area. This must be particularly the case for relatively localized areas of growth since these will be more open to both labour migration and commuting flows; hence general levels of demand deficiency in the surrounding region will tend to be more important than all but the most substantial employment changes within the area.

Patterns of Unemployment in Areas of Growth and Decline

In the light of these theoretical predictions as to how unemployment rates should be affected by local employment growth, this section explores the

actual pattern of unemployment in the growth areas and compares it with
the situation in the areas of most acute employment decline—Greater London and the other conurbations. Most of this section is based on a reanalysis
of data from the 1981 LFS which allows the calculation of unemployment
rates for precisely defined groups, and also contains data on unemployment
duration and changes in employment status over the previous year. However, some use has also been made of data from the 1981 Population Census
which was taken during the period of LFS sampling, since the number of
individuals covered by its analyses is much larger.

Table 4.4 shows how unemployment rates in the growth areas compared
with those elsewhere in the country in 1981, distinguishing a number of
groups who generally face a higher probability of unemployment. For the
growth areas as a whole, unemployment in each group was clearly below the
national average, and substantially below that in Inner London and in the
other conurbations, although the rates among unskilled manual men (socio-
economic group (SEG) 11) and the youngest group of women were apparently higher than for Greater London as a whole. Moreover, the differences
between unemployment rates in the growth areas and the conurbations are
greater for the disadvantaged groups than for all workers, so that the disparities in unemployment rates between different groups are lower in the growth
areas. For example, the overall unemployment rate for men is nearly 8 per
cent lower in the growth areas than in the conurbations, while the rate for
men without qualifications is about 10 per cent lower, as is the rate for men
aged 16–19. For non-white men the difference is 13 per cent while for
unskilled men (SEG 11) it is nearly 14 per cent. In the case of women, the
position in the growth areas is even more consistently favourable in comparison with the conurbations. Semi-skilled men (SEG 10) and men aged 55 and
over are the only groups whose relative position appears less favourable in
the growth areas in that their margin above average is greater than it is in the
conurbations. For both men and women, however, the relative advantage of
those in the growth areas is much less clear if the comparison is made with
Inner London, or with Greater London as a whole. For example, the unemployment rate for semi-skilled male manual workers in Inner London is 3 per
cent less than the rate for all males, while in the growth areas it is 1 per cent
higher. Among unskilled male workers the rate is higher than the average in
both areas, but by 3 per cent in Inner London and nearly 9 per cent in the
growth areas. Nevertheless, the general pattern is for unemployment rates
among potentially disadvantaged groups in the growth areas to compare
very favourably with those elsewhere, including average rates for Britain as a
whole.

However, it is also notable that there are substantial differences between
the growth areas. The most outstanding is the much higher level of unemployment in the two New Towns groups. This is the case in southern New
Towns for each of the population groups which we have distinguished,

Table 4.4. *Unemployment Rates (%) for Disadvantaged Groups, 1981*

	No qualifications (LFS)	Age 16-17 (LFS)	Age 16-19 (Census)	Age 20-4 (Census)	Age 55+ (Census)	SEG 10 (Census)	SEG 11 (Census)	Non-white (LFS)	All males (Census)
Males									
City regions	8.9	19.3	15.3	12.9	9.8	9.1	17.9	7.1	8.7
Southern New Towns	12.3	28.6	18.1	15.6	11.7	10.8	19.0	19.2	11.0
Northern New Towns	21.2	30.8	22.5	19.4	16.4	14.2	24.4	38.5	13.2
Urban areas	10.2	27.3	12.9	10.9	8.8	9.1	16.4	5.7	7.3
Rural areas	7.4	17.4	13.0	10.5	10.0	8.7	14.4	5.7	7.3
Metropolitan fringe	6.6	21.4	13.2	10.0	7.8	7.4	16.2	6.8	6.4
All growth areas	9.3	22.7	14.1	11.6	9.5	9.0	16.5	8.4	7.8
Inner London	14.6	34.8	25.2	18.0	11.7	10.7	17.1	14.7	14.0
Greater London^a	11.2	28.0	19.2	14.0	9.5	9.1	15.1	10.8	10.1
Other Conurbations	18.9	33.3	24.4	21.7	16.9	16.3	30.1	21.4	15.5
Great Britain	13.7	28.8	18.6	15.9	12.3	12.2	22.9	13.6	10.9

	No qualifications (LFS)	Age 16-17 (LFS)	Age 16-19 (Census)	Age 20-4 (Census)	SEG 10 (Census)	Non-married (Census)	Non-white (LFS)	All females (Census)
Females								
City regions	4.7	15.8	14.0	9.1	6.0	10.4	8.3	5.9
Southern New Towns	11.0	30.4	15.5	10.3	7.2	11.8	26.3	7.0
Northern New Towns	13.0	37.5	20.9	14.1	10.3	15.2	16.7	8.6
Urban areas	5.8	26.3	12.4	8.5	6.5	9.6	11.4	5.3
Rural areas	7.1	21.6	12.3	8.3	5.2	9.7	8.3	5.5
Metropolitan fringe	4.9	19.0	9.7	6.3	4.8	7.3	2.3	4.4
All growth areas	6.5	21.9	12.7	8.4	5.9	9.6	8.4	5.5
Inner London	9.4	24.6	19.9	11.4	8.6	11.9	9.6	8.9
Greater London	7.9	18.0	14.9	9.2	6.7	9.9	9.4	6.7
Other conurbations	11.4	37.3	21.9	14.6	10.6	15.5	15.1	9.1
Great Britain	9.2	27.8	16.7	11.1	8.1	11.9	11.0	7.1

^a Including parts of 'metropolitan fringe' within Greater London.

Source: 1981 LFS and Census of Population.

though the margin is not large for semi-skilled workers, and even more so in
the northern New Towns, where unemployment rates are comparable with
those in the conurbations. In the southern New Towns the main explanation
is to be found in the rapid population growth, to which we have already
referred. In view of the fact that these New Towns had the fastest overall
growth in employment, with significant increases even in manual jobs, it
shows clearly that the promotion of local growth does not necessarily lower
unemployment if it serves as much to encourage in-migration of people as of
jobs. Population growth was also a factor in the case of the northern New
Towns, but more important here were the high rates of unemployment in the
surrounding areas where there has been a substantial fall in employment.
Unemployment rates also tended to be somewhat higher in the city regions
than in the growth areas as a whole, presumably reflecting their rather lower
rates of employment growth, although the unqualified, black workers, and
new entrants to the labour force all tended to fare rather better there. Conversely, the youngest workers appear to do relatively badly in the two groups
of New Towns and the 'urban areas', in each case probably as a result of the
larger supply of young workers in areas with higher rates of natural increase.
This factor may also account for the relatively high unemployment among
black workers in New Towns, since high proportions of these workers are
young.

Table 4.5 looks at the probabilities of moving into unemployment over the
course of a year for the main occupational groups in the different areas. It
derives from the 1981 LFS and shows the percentage unemployed at the time
of the survey according to occupational group a year previously, as well as
the percentage of those unemployed a year ago who were unemployed at the
time of survey (although not necessarily throughout the intervening period).
In all areas manual workers were much the most likely to move into unemployment, and there tended to be a steady gradient from APT to manual
worker in unemployment probabilities. This gradient was much less steep in
the growth areas than in the conurbations or Inner London, indicating that
service and manual workers, who are all potentially disadvantaged, gain particularly from the lower risks of unemployment in the growth areas.

For those who become unemployed, the probability of remaining out of
work (or of incurring a second spell within the year), and thus expected
duration of unemployment, also appears significantly lower in the growth
areas than in Inner London or the other conurbations. Inner London's position, however, reflects particularly its concentration of unskilled and single
men who have lower probabilities of leaving unemployment, and, for men at
least, expected durations in the growth areas are very similar to those in
Greater London as a whole.

A more direct measure of unemployment duration is presented in Table
4.6, although the proportions of long-term unemployed recorded here are
depressed by the rapid rise in unemployment in the year before the survey.

Table 4.5. *Percentage Unemployed in 1981 by Occupation in Previous Year*

	APT	Clerical	Service	Manual	Unemployed
Males					
City regions	2.6	2.2	4.7	4.6	60.0
Southern New Towns	3.7	2.1	3.9	9.7	45.8
Northern New Towns	6.0	6.9	9.1	13.9	72.7
Urban areas	2.1	3.5	3.2	5.3	57.5
Rural areas	1.9	4.0	4.0	4.7	61.2
Metropolitan fringe	1.4	2.4	3.8	5.0	45.6
All growth areas	2.1	3.2	4.3	5.8	57.8
Inner London	2.5	3.4	7.1	9.0	61.5
Greater London	2.3	2.2	5.2	7.5	56.9
Other conurbations	2.9	3.5	7.2	10.3	67.8
Females					
City regions	2.0	3.1	2.3	7.9	21.2
Southern New Towns	1.3	4.4	5.8	12.0	66.7
Northern New Towns	4.7	6.3	6.2	12.7	60.9
Urban areas	2.5	2.4	2.9	6.4	38.2
Rural areas	2.8	2.5	3.8	3.1	50.0
Metropolitan fringe	1.6	2.2	2.6	6.2	28.6
All growth areas	2.3	2.8	3.3	6.3	42.7
Inner London	3.2	4.6	4.8	6.5	50.9
Greater London	2.7	2.8	3.4	5.8	50.0
Other conurbations	2.0	3.5	3.6	8.4	60.3

Source: 1981 LFS.

The proportion of long-term unemployed in the growth areas is significantly lower than that in the other conurbations, and this holds for almost all sub-groups. Given the lower overall level of unemployment, this means that the long-term unemployed were a very much lower proportion of the economically active population (1.8 per cent for males as compared with 5.5 per cent in the 'other conurbations'). There was much less difference from the proportion of long-term unemployed in London, except in the cases of young and non-white men.

Tables 4.5 and 4.6 together point to some significant differences between the growth areas in likely durations of unemployment. For males these appear to be inversely related to the rate of employment growth, while for females they simply reflect the level of unemployment. The northern New Towns come out badly on either count, but in the southern New Towns the combination of fast employment and population growth seems to have produced large flows into unemployment for men but shorter durations.[8]

This research has focused on unemployment as its indicator of labour-market disadvantage, both because it is the most acute manifestation of such disadvantage, and because it is much the easiest to measure. It is not, how-

Table 4.6. *Percentage of Unemployed Seeking Work for more than Twelve Months, 1981*

	No qualifications	Age 16–20	Age 21–4	Age 55+	SEG 10	SEG 11	Non-white	All males
Males								
City regions	(36.1)	(31.7)	(18.8)	(33.3)	(28.0)	(44.4)	(20.0)	29.2
Southern New Towns	(13.3)	(9.1)	(0.0)	(33.3)	(25.0)	(20.0)	(20.0)	10.7
Northern New Towns	(30.6)	(19.0)	(30.0)	(40.0)	(20.8)	(11.1)	(20.0)	27.1
Urban areas	33.3	13.2	(27.8)	(40.5)	(25.8)	(50.0)	(25.0)	22.6
Rural areas	(30.3)	10.0	(17.4)	(55.6)	(23.3)	(40.0)	(50.0)	26.6
Metropolitan fringe	(24.1)	(11.9)	(12.5)	(29.2)	(0.0)	(31.2)	(11.1)	16.5
All growth areas	29.8	16.0	18.3	40.9	21.9	34.9	21.9	23.2
London	32.7	23.5	29.8	39.2	20.3	35.2	36.7	27.6
Other conurbations	44.4	29.1	36.3	37.9	36.8	43.6	32.3	35.2

	No qualifications	Age 16–20	Age 21–4	Age 55+	SEG 10		Non-white	All females
Females								
City regions	(19.4)	(7.4)	(0.0)	(33.3)	(12.5)		(25.0)	9.6
Southern New Towns	(16.7)	(26.7)	(0.0)	(33.3)	(16.7)		(20.0)	16.4
Northern New Towns	(25.0)	(17.6)	(33.3)	(0.0)	(36.4)		(0.0)	(24.1)
Urban areas	12.5	10.3	(13.3)	(14.3)	(7.1)		(40.0)	14.2
Rural areas	(34.9)	7.9	(26.7)	(55.6)	(33.3)		(66.6)	21.8
Metropolitan fringe	(17.6)	(3.1)	(0.0)	(22.2)	(7.1)		(50.0)	12.4
All growth areas	22.4	10.1	12.2	27.3	20.3		35.0	16.3
London	19.3	14.3	14.8	22.2	18.2		20.0	16.4
Other conurbations	34.5	27.0	26.0	42.4	24.3		24.5	28.3

Note: Figures in parentheses have standard errors greater than 5%.

Source: 1981 LFS.

ever, the only form of disadvantage. A later section will consider evidence on employment instability, but this section concludes by considering evidence on wage variation. Satisfactory data are not available for the individual growth areas, but some indications of relativities can be derived from data for the South East outside Greater London, East Anglia, and the South West. In all cases for women, and in all but the first for men, money wages are below the national average, but this may partly reflect differences in living costs. On the assumption that these differences are fully reflected in the wages of the more mobile non-manual groups, a measure of degrees of disadvantage can be derived from a comparison of the wages of manual and service workers relative to these groups. In London these show a marked pattern of disadvantage, since the groups with the lowest wages in absolute terms also have the lowest relative wages, and by implication the least compensation for higher living costs. In the growth regions, on the other hand, there is no evidence of particular disadvantage. Manual wages for men and women in each of the regions tend to be higher relative to the national rates than are non-manual wages. For men this also holds for the lowest-paid occupational groups: service workers, miscellaneous workers (mainly labourers), and assembly workers. The pattern is less clear-cut for women, but there is no evidence of the pattern of disadvantage observable in London.

Examination of aggregate patterns of unemployment shows clearly that disadvantaged groups in the work-force are much less likely to be unemployed if they live in areas of employment growth rather than employment decline, and that their chances of employment are much more dependent on the occurrence of growth than are those of other workers. However, the fact that there is considerable variation in this respect between different types of growth areas, and that disadvantaged groups appear to suffer less in London, despite its chronic employment decline, than in other conurbations, indicates that local employment growth is not the only factor influencing their exposure to unemployment. Employment change within the wider region is also important, while at the local level it is the relationship between rates of employment and population change which is most relevant. Taking these qualifications as given, evidence on both unemployment durations and wage levels indicates that the benefits to disadvantaged groups living in an area of employment growth extend beyond the simple fact of lower unemployment rates. Indeed, in the case of unemployment durations for male workers there appear to be benefits from local employment growth even where population is also growing.

Individuals' Characteristics and the Incidence of Unemployment

Disadvantage in the labour market is commonly the outcome of a combination of characteristics of an individual—gender, race, age, qualifications, work history, and so on—which interact with each other.

The extent to which particular combinations of characteristics serve to disadvantage individuals in different sorts of labour markets can best be seen by comparative analyses of individuals' employment status as recorded in the LFS. The method which we have adopted for this purpose is to estimate for each group of areas a set of logit functions, a form of regression analysis appropriate to discrete variables such as employment status, which records a person as either being or not being unemployed at a particular point in time. These functions relate the probability of being unemployed to a series of personal, household, or employment characteristics which are assumed to combine multiplicatively to determine the relative odds of being unemployed rather than employed. The estimates might, for example, suggest that the relative odds for young workers and for the unqualified are each three times as great as for, respectively, older workers and those with qualifications; in this case the relative odds for workers who are *both* unqualified *and* young would be nine times those for older qualified workers. Logit analyses of this type were carried out separately for males and for females in the growth areas (taken as a single group) and, for comparison, in Greater London and in the metropolitan counties (including Central Clydeside). These three sets of areas include contrasts both between growing and declining parts of southern England (since most of our set of growth areas are in the South) and between regions of relatively higher and lower unemployment.

The definition of 'unemployment' adopted for these analyses was intended to distinguish those who were available for work at the time of the survey in 1981 but not in employment, irrespective of whether they were actively seeking work or registered as unemployed. It thus includes both those recorded as 'seeking work' and a proportion of the 'other economically inactive', excluding only students, housewives, the retired, and the long-term sick or disabled. About three-quarters of the male unemployed on this definition were actively seeking work, while this was the case for about half of the female unemployed. Those at risk of being unemployed were assumed to be the unemployed themselves (as defined here) plus those currently in employment.

The estimated logit functions are presented in Tables 4.7 and 4.8. Comparison of the constant terms shows that for workers with a standardized set of characteristics[9] the probability of unemployment was rather lower in the growth areas than in most of the declining metropolitan areas. For males with these characteristics, however, the probability of unemployment appeared to be as low in Greater London as in the growth areas, the main division in this case being between growing and declining groups of regions, rather than a sub-regional division, whereas the latter appeared more important for females. These comparisons are a little unrealistic since the growth and stability characteristics of industries in the growth areas are not identical to those in the conurbations. If we allow for this by using average characteristics for industries in each of the groups of areas as a bench-mark, the relative position of the growth areas appears even more favourable. The 'standard male' in the growth areas would then appear to have about 20 per cent less chance of being unemployed than if he lived in Greater London and about 70 per cent less than if he lived in one of the other metropolitan areas; for the 'standard female' the probability of unemployment in the growth areas would be about 35 per cent less than in London and about 60 per cent less than in the other conurbations. These differences in relative odds of unemployment as against employment are actually rather larger than the overall differences in unemployment rates, implying (though this is only a partial basis for comparison)[10] that the lower unemployment rates in the growth areas are not explicable in terms of the particular characteristics of their labour force. By contrast, the London study (Buck *et al.* 1986) found that variations in unemployment rates within London—particularly the disparity between Inner and Outer London—were substantially associated with the competitive position of groups resident in different parts of the city. The lower unemployment in the labour markets of the growth areas does, however, seem to be primarily a reflection of more favourable employment changes (relative to population trends), and possibly a greater stability in employment conditions.

The significant coefficients on many of the variables included in the logit functions, however, show that there are very great disparities in the chances of being unemployed between different groups within the growth areas. Among males, 16–24 year-olds and the over-60s,[11] the unmarried, disabled or black workers, public sector or housing association tenants, those without identifiable formal qualifications, and those in unskilled or unclassified jobs all appear to have much greater than average chances of being unemployed. In the case of black workers (and the disabled), for example, the relative odds of unemployment were $2\frac{1}{2}$ times as great; for both the 'unskilled' and the unqualified they were about 80 per cent greater. And these sources of disadvantage are cumulative, so that for an unqualified and unmarried black worker, aged 16–19, in an unskilled manual occupation, living in a local authority tenure, the relative odds of unemployment as against employment appear to be about *30 times* as great as for an individual with the standard characteristics. Among females, youth, disablement, Asian or 'mixed' ethnic origin, lack of qualifications, and in this case *semi-skilled* manual employment are similarly disadvantaging characteristics. For both males and females, the characteristic most clearly associated with unemployment is having an unclassifiable occupation (in the last or current job); this position may be both cause and consequence of exposure to unemployment, but remains highly significant if new entrants or re-entrants to the labour force are excluded. The form of the analysis does not allow a direct assessment of whether gender itself affects the chances of unemployment, but we may note that on our definition of unemployment female and male rates were very similar in the growth areas whereas male unemployment was somewhat

Table 4.7. *Logit Estimates for Probability of Males being Unemployed, 1981*

Characteristic	Growth areas		Greater London		Other metropolitan areas	
Age: 16–19	0.477	(2.7)	0.416	(2.0)	0.043	(0.4)
20–4	0.312	(2.0)	0.320	(1.9)	0.388	(3.9)
25–9	—		—		—	
30–49	−0.190	(1.5)	−0.394	(2.7)	−0.340	(3.9)
50–4	−0.177	(1.0)	−0.294	(1.5)	−0.538	(4.3)
55–9	−0.152	(0.8)	−0.864	(3.7)	−0.423	(3.3)
60–4	0.802	(4.6)	0.293	(1.3)	0.334	(2.6)
Married	−0.518	(4.4)	−0.377	(3.0)	−0.492	(5.7)
Living with parents	−0.083	(0.5)	−0.116	(0.7)	−0.234	(2.2)
Registered disabled	1.054	(5.4)	1.088	(4.4)	0.630	(3.9)
Afro-Carribean ethnic origin	0.856	(2.1)	1.009	(6.1)	0.507	(2.5)
Asian ethnic origin	0.932	(2.9)	0.900	(4.9)	0.922	(6.4)
Mixed or 'other' origin	0.522	(1.1)	0.583	(2.0)	0.436	(1.3)
Tenure: owned outright	0.493	(3.1)	0.127	(0.7)	−0.354	(2.7)
mortgagee	0.013	(0.1)	−0.354	(2.3)	−0.444	(3.7)
council tenant	0.818	(5.5)	0.691	(4.8)	0.673	(6.0)
New Town tenant	1.642	(8.0)	—		0.224	(0.7)
Housing Association	1.000	(3.0)	0.828	(3.5)	0.796	(3.6)
private rented	—		—		—	
Qualification: degree	0.037	(0.2)	0.156	(0.7)	−0.436	(1.8)
unspecified	0.976	(5.0)	0.621	(2.1)	0.844	(4.9)
none	0.614	(7.0)	0.827	(7.4)	0.550	(9.0)
SEG: personal service	0.388	(1.2)	−0.562	(1.6)	0.643	(2.7)
unskilled manual	0.602	(4.5)	0.494	(2.9)	0.684	(8.2)
unclassified	3.494	(29.5)	2.952	(23.0)	4.124	(34.4)
Industry: employment stability (10%)	−0.805	(6.8)	−0.894	(6.7)	−0.777	(10.5)
employment growth (10%)	−0.632	(9.0)	−0.713	(7.7)	−0.504	(13.1)
Constant	−4.522		−4.580		−3.517	
No. of unemployed in sample (and rate)	987	(7.4%)	656	(8.9%)	2350	(18.2%)

Notes: 1. The dependent variable includes those seeking work plus the 'other economically inactive' excluding the permanently sick and disabled.

2. Coefficients are on a logarithmic basis: the exponential of any value represents the proportional change in the relative odds of unemployment against employment associated with that characteristic. Values in parentheses in the main part of the table are asymptotic t statistics.

3. The employment stability variable was based on the number of individuals in each industry and area type remaining with the same firm during the 12 months preceding the survey, expressed as a proportion of the average numbers employed at the time of the survey and 12 months previously: the mean values were 86% for males and 82% for females with standard deviations of 4%. The employment growth variable was based on the change in numbers employed in each industry and area type in the 12 months preceding the LFS: the mean values were −3% for males and +1% for females with standard deviations of around 7%. Zero values for the two variables represent perfect stability in employment and no change in numbers. Values of −1 represent 90% stability and a decrease of 10% in employment.

Source: 1981 LFS.

Table 4.8. *Logit Estimates for Probability of Females being Unemployed, 1981*

Characteristic	Growth areas		Greater London		Other metropolitan areas	
Age: 16–19	0.917	(3.7)	1.058	(3.8)	0.498	(2.7)
20–4	0.618	(3.0)	0.247	(1.1)	0.424	(2.7)
25–9	—		—		—	
30–49	−0.340	(1.9)	−0.426	(2.1)	−0.877	(6.3)
50–4	−0.145	(0.6)	−0.605	(2.1)	−0.760	(4.0)
55–9	−0.134	(0.5)	−0.486	(1.7)	−0.908	(4.6)
60–4	0.448	(0.1)	−0.210	(0.5)	−1.386	(3.3)
Married	−0.034	(0.2)	−0.500	(3.3)	−0.566	(4.8)
Living with parents	−0.370	(1.7)	−0.370	(1.6)	−0.750	(4.9)
Registered disabled	1.090	(2.8)	1.343	(2.8)	0.539	(1.5)
Afro-Carribean ethnic origin	0.358	(0.6)	0.913	(4.3)	0.376	(1.1)
Asian ethnic origin	1.105	(2.2)	1.060	(4.2)	0.644	(2.1)
Mixed or 'other' origin	1.245	(2.5)	0.716	(1.9)	−0.038	(0.1)
Tenure: owned outright	0.570	(2.5)	0.212	(0.9)	0.464	(2.0)
mortgagee	0.430	(2.2)	0.076	(0.4)	0.644	(3.1)
council tenant	0.629	(3.0)	0.284	(1.5)	1.080	(5.2)
New Town tenant	2.026	(7.2)	—		0.356	(0.6)
Housing Association	1.294	(3.1)	0.599	(1.8)	1.192	(3.5)
private rented	—		—		—	
Qualification: degree	0.412	(1.3)	0.720	(2.7)	0.431	(1.2)
unspecified	0.784	(3.5)	0.753	(2.3)	0.385	(1.6)
none	0.445	(3.5)	1.304	(8.3)	0.809	(7.5)
SEG: personal service	0.324	(1.8)	−0.165	(0.7)	−0.138	(0.9)
semi-skilled manual	0.726	(4.8)	0.168	(0.7)	0.190	(1.6)
unskilled manual	−0.501	(1.7)	−1.280	(2.5)	−0.300	(1.6)
unclassified	4.720	(31.2)	3.822	(24.4)	5.392	(32.7)
Industry: employment stability (10%)	−0.525	(4.2)	−0.337	(2.2)	−0.726	(6.3)
employment growth (10%)	−0.307	(4.8)	−0.230	(2.6)	−0.412	(6.9)
Constant	−5.058		−4.360		−4.549	
No. of unemployed in sample (and rate)	670	(7.6%)	448	(8.3%)	1185	(11.3%)

Notes: See Table 4.7.

Source: 1981 LFS.

higher in Greater London and much higher in the metropolitan areas of the North and Midlands.

Not all of the 'characteristics' included in these analyses can be regarded as independent of the labour-market position which the individual occupies (see Norris 1978). For example, having a last occupation in an 'unskilled manual' job, a position occupied by some $4\frac{1}{2}$ per cent of men in the growth areas, is by no means the same thing as lacking specifiable skills. This is an issue to which we shall return in the next section. An example which emerges more clearly from the logit analyses is the importance of the industry of employment, even when socio-economic status has been controlled for. For both males and females in the growth areas, rates of employment decline and turnover in the industry within which individuals were last employed appear among the most important determinants of the chances of unemployment.

Comparison of the estimated logit functions for the samples of individuals in the growth areas and in the two conurbation groups suggests that the patterns of advantage and disadvantage are essentially similar and that the disparities in chances of unemployment are broadly comparable in the three area types. Since the unemployment rates are quite different in the three samples, this lends some further support to the hypothesis that disadvantages tend to have proportional rather than constant impacts on the incidence of unemployment—and hence that employment growth or decline within an area has most effect on unemployment levels among those groups in the weakest competitive position.

In terms of the effects of specific characteristics, few significant differences can be detected between the growth areas and the conurbations, despite the size of the samples available from the LFS. There are a number of apparent differences between age groups. By comparison with the metropolitan areas, but not London, unemployment appears relatively high among 16–19 year-olds—perhaps implying that the disadvantage experienced by new entrants to the labour force may not actually be proportional to the level of unemployment. The age groups which appear to benefit most from the relatively strong labour markets of the growth areas are those in their 20s, with the over-50s in particular having higher unemployment rates than might have been expected. It would seem that perhaps seniority is more of an advantage in retaining employment in areas of decline than it is in gaining it where employment growth has been occurring. In the case of both the disabled and black workers, there appear to be some significant differences between the two 'southern' samples and the metropolitan areas—with relative disadvantage being less in the latter—rather than between the areas of growth and decline.

The largest of the groups experiencing an evident disadvantage in terms of the chances of being unemployed are those without formal educational, professional, or trade qualifications including one in three of the workers in the growth areas. Men in this category appear similarly disadvantaged in the growth areas as in the conurbations, but unqualified women do seem to be more protected from the risks of unemployment in the growth areas. However, both personal service workers and semi-skilled manual workers—the two groups of women with the highest incidence of unemployment—appear to do relatively worse in the growth areas; among males the only significant occupational difference between the samples is that personal service workers appear to have a lower incidence of unemployment in London, possibly because the concentration of service jobs there shortens the durations of unemployment among this high-turnover category of workers.

In terms of housing tenures, among males mortgagees are generally the most favoured group—a position attributed by McCormick (1983) to the greater disincentive for them to remain unemployed under current benefit structures—but in the growth areas they share this position with private renters and in the northern conurbations with outright owners. Among females it is private renters who are generally the most favoured group. In the London study (Buck *et al.* 1986) we have argued that these variations probably have more to do with aspects of social class affecting employers' perceptions of various groups of workers than with housing *per se*. Certainly no simple explanation adequately covers the variation between these three groups of areas. However, we may note that it is generally the public sector tenants who seem to do worst and that, relative to mortgagees in particular, their degree of disadvantage does not appear much different as between the growth areas and the conurbations. The one group who do seem to have a particularly great risk of unemployment in the growth areas are New Town tenants, but more detailed analyses make it clear that this reflects the higher unemployment generally in New Towns rather than a tenure difference within these towns.

In summary, the result of these analyses is to confirm the hypothesis that disadvantaged groups should benefit *proportionately* from living in areas of employment growth rather than decline, and thus experience a larger absolute reduction in their unemployment rate than other groups of workers; proportionate disparities in the incidence of unemployment appear as great in the growth areas, however, and there is little evidence of specific advantages for any particular groups as a consequence of local employment growth. Labour-market disadvantage appears to occur in a very similar form in both growing and declining areas, although it is more acute in the latter.

Labour-market Influences on Access to Stable Employment

The analysis in the two preceding sections takes a relatively short-term view of the possible relationship between employment growth and unemployment, treating a worker's occupation, the stability of jobs in the industry of last employment, and his or her qualification levels as independent influences on the risks of being unemployed. In contrast to this assumption, a segmented labour-markets perspective would treat such characteristics of the labour-market position occupied by workers as contingent upon the general structure and performance of the particular labour market and upon the workers' past experience of employment change or stability. Norris (1978) and Cousins and Curran (1982) show that even workers with past employment in skilled work are liable to get restricted to 'unskilled' jobs in unstable sectors (or 'secondary employment' as it is often called) as a *consequence* of involuntary unemployment. Such studies emphasize the particular difficulty for unemployed people of securing stable jobs in organizations which have developed internal labour markets (that is, primary sector jobs), particularly in circumstances of high unemployment. We might expect, therefore, that declining areas would tend to exhibit an increasing proportion of 'sub-employed' workers with periodic spells of unemployment and diminishing prospects of competing successfully for available jobs against other potential workers including migrants from outside the area. In growth areas, or areas of low unemployment, we should expect that fewer of those who are potentially disadvantaged—for example, by lack of qualifications—would be in this particularly vulnerable position and that more would be in secure jobs.

This hypothesis implies that in an area of high unemployment or employment decline there should be an increasing proportion of unskilled or secondary sector workers, independent of any changes in the number of secondary sector jobs. This might particularly be the case for males, whose employment opportunities are more clearly stratified (as between primary and secondary employment), only a small minority of (mostly well-qualified) women being in stable primary jobs. Tests of the hypothesis are complicated by the extent of migration between areas, since selective migration out of areas of decline and unemployment is liable to alter the occupational structure of their workforce, even if experience of unemployment or redundancy does not affect individuals' occupational position. The need to control for this factor limits possible tests to the period 1966–71 for which (uniquely) it is possible to link Census data on changes in the number of economically active males by SEG and on migration within Great Britain. Even for this period, data are lacking on overseas emigrants so it has been necessary to ignore the impact of international migration. Nevertheless, by deducting the number of net migrants in each SEG (as at the end of the period) from the overall change for an area during the period, it is possible to derive an approximation to the occupational or status change experienced by those resident in the area at the start of the period (including those who subsequently moved elsewhere).

Estimates of the implied occupational shift as it would have affected numbers of unskilled male manual workers in each region and conurbation are shown in Table 4.9. It can be seen that this group declined in almost all areas between 1966 and 1971 (the exception being the Tyneside conurbation) but that there were substantial variations in the rate of decline. The areas of slowest decline were in regions with unfavourable employment trends, including the Northern region, Yorkshire and Humberside, Wales, and to a lesser extent Scotland, although *not* the North West. At the sub-regional level, however, there was no tendency for the areas with the least favourable employment changes (that is the conurbations) to experience a slower shift out of unskilled jobs; indeed Greater London shows the second-fastest shift, despite a great shrinkage in employment. The clearer link is with levels of unemployment at the start of the period, which was above average in the northern and western regions. However, the numbers of unskilled in Yorkshire and Humberside remained higher, and in the North West (particularly Merseyside) fell faster, than might have been expected on this basis.

These variations may reflect the effects of either differential patterns of industrial change in the two regions, or perhaps international migration. However, for this period at least there does seem to be some evidence that the general shift out of unskilled jobs was slower in areas with worse employment experience. Levels of unemployment rather than employment decline as such seem to have been the critical factor: substantial intra-regional disparities in rates of employment change, balanced by population decentralization, are not reflected in differing rates of movement out of the unskilled category; on the other hand, inter-regional inequalities in unemployment do appear to have a direct effect on individuals' chances of being confined to this section of the labour market, with consequently greater exposure to unemployment.

As Table 4.9 also shows, net migration offset only a small proportion (less than 10 per cent) of the differences between areas in rates of change in the numbers of unskilled manual workers, compared with about three-quarters of the differences for other groups of male workers. Selective migration (or selective non-migration) thus boosted the tendency for the declining areas to acquire relatively higher proportions of unskilled workers. Unless the relative decline of manual employment was also proceeding more slowly in those areas, which does not seem to have been the case, the consequence must have been an increasing imbalance between the demand and supply for unskilled workers. In terms of the conventional classification of types of unemployment (as used, for example by Dixon and Thirlwall 1975) the demand-deficient unemployment created initially by employment decline would thus be converted into structural unemployment.[12]

Although this analysis has had to be undertaken with data for a period well before the current recession, its main results are of some significance at the present time. In particular, it suggests that there are some important differences between areas in the tendency for sub-employment to increase or decrease. In the late 1960s, to which our data refer, these areal variations were around a general contraction in the numbers of men in the unskilled, and potentially sub-employed, category. In the 1980s, with an excess supply of workers for the available jobs in any area, the numbers of unskilled and sub-employed would tend to be increasing everywhere. Once again, the important differentiating characteristic of areas appears to be the rate of unemployment rather than of employment growth *per se*, with the clearest distinction in the observed shifts out of unskilled manual work being between the North and the South. It is particularly notable that in this case Greater London appears as favoured as anywhere else in the South. Finally, the analysis confirms that inter-regional migration has been highly socially selective, contributing very little to adjustments to imbalances in the demand and supply for unskilled manual workers, with the unskilled workers left behind forming an increasing proportion of the labour force in areas of employment decline.

A downward spiral of occupational mobility is one of the ways in which

Table 4.9. *Occupational Shifts by Region and Conurbation, 1966–1971*

Area	% change in male economically active 1966–71	% change less effect of domestic migration	% change in unskilled male manuals 1966–71	% change less effect of domestic migration	Male unemployment rate 1966 (%)
Northern region	-2.0	-1.0	-1.0	-1.0	3.8
Tyneside conurbation	-5.0	+0.5	+2.2	+3.3	4.2
Yorks. and Humberside	-1.4	-0.1	-2.7	-2.6	2.2
W. Yorks. conurbation	-3.2	-0.8	-0.7	+0.2	2.0
North West	-1.9	-1.4	-7.9	-8.6	2.9
SELNEC[a]	-4.5	-1.8	-8.1	-9.0	2.5
Merseyside conurbation	-8.4	-0.5	-14.6	-10.2	4.4
E. Midlands	-1.1	-2.3	-9.3	-9.9	1.8
W. Midlands	-0.6	+0.1	-10.0	-10.0	1.8
W. Midlands conurbation	-4.7	+0.7	-12.1	-10.9	1.8
E. Anglia	+4.2	-0.3	-10.5	-12.6	2.4
South East	+0.2	+0.1	-13.6	-12.9	2.1
Greater London	-7.0	-1.1	-15.9	-13.3	2.3
South West	+1.7	-1.3	-12.5	-14.0	2.5
Wales	-1.5	-1.4	-1.0	-1.1	4.2
Scotland	-1.8	-0.6	-5.4	-4.3	4.4
Clydeside conurbation	-4.8	-0.1	-7.9	-5.1	5.5
Great Britain	-0.7	-0.7	-8.4	-8.4	2.6
Standard deviation (unweighted)	4.1	1.3	5.6	5.2	

[a] SELNEC, South-east Lancashire and North-east Cheshire.

Source: 1966 and 1971 Censuses of Population.

individuals may acquire disadvantaging characteristics as a consequence of working in a declining labour market. However, as we have shown in the previous section, it is not only unskilled manual workers who face particular risks of unemployment but also those working in industries characterized by above-average rates of employment turnover. As segmented labour-markets theory implies, jobs vary systematically in the extent to which they promote or inhibit employment stability. Analysis of LFS data on the probability of individuals remaining with the same firm during a twelve-month period shows that there are significant variations on both occupational and industrial dimensions which are not simply reflections of characteristics of the workers involved (such as age, marital status, or qualifications levels). For males there were a small minority, and for females a large majority, of occupations (SEGs in fact), thought of as less skilled, in which individuals were particularly liable to change firms. Holding this factor constant, there were also consistent industrial differences for men and for women, apparently reflecting the degree of competition/ monopoly in the relevant product markets. Relatively stable jobs were found in the more concentrated manufacturing sectors, public services, and financial services, while the least stable were in construction and a range of other private services. The occupational and industrial dimensions of stability interacted, however, with employers, managers, and professionals being unaffected by the industry in which they worked, and the unskilled or unclassified proportionately much more so.

Using the results of this analysis, a comparison was made of the stability attributes of the industry/occupational structure in the growth areas as compared with the conurbations, with data again from the 1981 LFS. The main question on which this comparison focused was whether there were detectable differences in the conditions of access to the more stable jobs. Because of the lack of commensurability between the occupational classifications of men's and women's jobs, the comparison was limited to male employment. Initially, separate analyses of variance were undertaken for each of the three samples (for the growth areas, Greater London, and the metropolitan countries), but these yielded extremely similar results, differing only in the significance of the ethnic factor, because of the limited numbers of black workers in the growth areas. No significant differences were evident in the conditioning effect of age, marital status, ethnic origin, or qualifications levels on access to stable jobs. When the three samples were pooled, a significant difference was evident between the sets of areas, but it was principally between the northern and southern groups rather than between the growth areas and the conurbations (see Table 4.10). Indeed, when the effects of qualifications and ethnic origins in particular were controlled for, it was those living in Greater London who appeared most likely to occupy a stable job. Within particular industries and occupations, rates of turnover were actually rather lower in the growth areas than in the conurbations, but this would seem to reflect the

Table 4.10. *Analysis of Variance in the Instability Characteristics of Jobs*

Source of variation	Unadjusted deviation	Adjusted deviation	Mean sum of squares	F
Age:			0.916	180.6
16–19	+0.041	+0.028		
20–4	+0.020	+0.015		
25–9	+0.002	+0.008		
30 and over	−0.009	−0.008		
Ethnic origin:			0.112	22.1
White	−0.001	−0.001		
Afro-Carribean	+0.031	+0.020		
Asian	+0.009	+0.011		
Mixed and other	+0.004	+0.007		
Marital status:			0.649	127.8
Married	−0.009	−0.004		
Not married	+0.019	+0.008		
Highest qualification:			5.206	1026.0
Degree, HNC, professional qualification	−0.062	−0.057		
ONC, A level	−0.030	−0.033		
Teaching, nursing, apprenticeship, O level	+0.001	+0.000		
CSE, other or still studying	+0.010	+0.001		
None	+0.022	+0.024		
Area of residence:			0.109	21.4
Growth areas	−0.005	−0.002		
Greater London	−0.004	−0.004		
Other conurbations metropolitan	+0.006	+0.003		
			$R^2 = 0.161$	

Note: Mean value, 0.115. Standard deviation, 0.077. All effects are significant at the 0.1% level.
Source: 1981 LFS.

differing behaviour of employers and workers in labour markets of differing scales, rather than the effects of differing rates of employment change.

These analyses of the conditions of access to stable jobs controlled for the effect of qualifications levels, which were clearly the most important influence (apart perhaps from the excluded question of gender) on whether individuals were in stable jobs. However, the structure and performance of local labour markets may have a very important bearing on the level of qualifications which are attained by individuals. This is most obviously the case for apprenticeships and other occupational qualifications, but Donnison (1980) has argued that educational aspirations and attainment in school, particularly among working class children, may also be substantially influenced by rates of employment growth in the local economy. Some support for this hypothesis was provided by evidence that, after controlling for family background, the performance of children in resort and engineering towns was significantly better than that of children from the conurbations or the older industrial cities.

In order to gauge the potential importance of such relationships a further set of analyses of variance was undertaken with the three area-based samples from the 1981 LFS. The dependent variable chosen in this case was the lack of formal qualifications—even the possession of a single Certificate of Secondary Education (CSE) pass—which was found to be the crucial distinction in terms of exposure to unemployment. The background factors which could be controlled for in this case were only age, ethnic origin, and marital status. No information was available on parental background or educational attainment, nor on the area of upbringing. The effect attributed to area of residence thus necessarily conflates any direct labour-market influences with those of social class composition and of selective migration. Nevertheless, it is noteworthy that in this case the area factor emerges as the most important of the independent variables, with the growth areas appearing as the most favoured of the three groups of areas (see Table 4.11). Once again, however, the contrast between the southern and northern samples appears sharper than that between the growth areas and the conurbations. In the conurbations of the Midlands and North (though not so evidently in London) there appears thus to be a cumulation of disadvantages linked to employment decline, with more workers lacking qualifications, more of these being likely to occupy unstable jobs, and a wider disparity in unemployment rates between those in stable and unstable employment.

Table 4.11. *Analysis of Variance for No Qualifications*

Source of variation	Unadjusted deviation	Adjusted deviation	Mean sum of squares	F
Age:			13.87	64.1
16–19	−0.06	−0.07		
20–4	−0.05	−0.06		
25–9	−0.05	−0.05		
30 and over	+0.02	+0.03		
Ethnic origin:			9.40	43.5
White	−0.01	−0.01		
Afro-Carribean	+0.17	+0.17		
Asian	+0.11	+0.11		
Mixed and other	+0.03	+0.04		
Marital status:			1.65	7.6
Married	−0.02	+0.01		
Not married	+0.01	−0.01		
Area of residence:			17.10	79.1
Growth areas	−0.04	−0.03		
Greater London	+0.00	−0.01		
Other conurbations/metropolitan	+0.03	+0.03		
			$R^2 = 0.014$	

Note: Mean value, 0.33. Standard deviation, 0.47. All effects are significant at the 0.1% level except for marital status which is significant at the 1% level.

Source: 1981 LFS.

The level of growth has significance too for the effectiveness of policies aimed at directly mitigating employment disadvantage. In particular, directly redistributive policies to eliminate discrimination are much more likely to be successful if new jobs are being created. Changes in public sector employment are particularly important, since it is here that anti-discrimination policy is most effectively implemented. In practice, however, recent employment growth in all areas has been skewed towards the private sector. In the case of ameliorative policies there is some evidence that these too are much more likely to work effectively in areas of lower unemployment, which is generally associated with employment growth. Greaves (1984) summarizes the results of a number of studies of the Youth Opportunities Programme (YOP) in contrasting labour markets. Although none of these labour markets were in growth areas, it appeared that in low-unemployment situations the programme did help to redress the employment disadvantage of groups such as West Indians, but notably failed to do so in high-unemployment areas. In these areas, where there was much more pressure on the programme as a whole, those with disadvantages such as low qualifications had to wait longer to enter the programme than the better-qualified, and there was a wider gap in the chances of obtaining employment after YOP.

Conclusions

The aim of this chapter was to examine the labour-market experience of disadvantaged groups in areas of employment growth, in order to contrast it with that found in declining urban areas and to see what inferences could be drawn about the effects of differing rates of local employment change on the position of these groups. Comparison of six main categories of growth area showed that employment change tended to be more favourable for all groups of workers and that in most areas this was reflected in distinctly lower unemployment rates and shorter durations of unemployment for all, including disadvantaged groups. Indeed, since relativities in unemployment rates tended to be proportional, these groups showed the largest reductions in probabilities of being out of work in areas where employment growth was occurring. Their relative earnings levels also appeared to be more favourable in these areas. Furthermore, it was found that there was more occupational mobility out of unskilled manual jobs in the areas with growing employment. Similarly, the proportions of workers in the more stable types of employment and of workers with some formal qualification were seen to be higher in the expanding areas. There is thus evidence for the view that the numbers of disadvantaged workers, as well as the severity of the disadvantages they face, are greater in areas (and presumably periods) of falling employment. Because of the greater mobility of other groups, there is a further tendency for the disadvantaged to become an increasing proportion of the labour force in areas where it is contracting.

Two qualifications to this general pattern were noted. First, as the relatively high unemployment rates in the expanding New Towns indicated, the advantages of employment growth in an area could be offset by any independent stimulus to population growth, although where both population and employment growth were occurring typical durations of male unemployment appeared shorter. Local unemployment rates could be as much affected by policies which stimulated or inhibited in or out-migration as by those bearing on employment change. Secondly, as the high unemployment rates in the northern New Towns particularly indicated, the effect of *local* employment growth on unemployment appeared quite limited, with much depending on rates of employment change and unemployment in the surrounding region. Most of the growth areas which we identified were concentrated in southern England, particularly in an arc from north-east of London round to the west and south of the city. Residents of these areas benefited from being within a broad region of favourable economic change, rather than merely from growth in their own areas. Indeed, the main contrasts which emerged in our analysis tended to be between North and South, rather than between the growing areas of the South and its declining conurbation.

Few significant differences were found in the *forms* of disadvantage as between any of the groups of growth areas, on the one hand, or the conurbations on the other. Rather, it appeared as though employment decline simply magnified disparities—as has happened generally in the current recession—with disadvantaged groups bearing the brunt of the adjustment process. As far as the differences between areas are concerned, *part* of the reason for this is to be found in constraints on mobility among manual and service workers, or the weaker links between employment change and the direction of movement among those outside primary labour markets. Hence there are larger disparities in unemployment rates between growing and declining areas for unskilled than for skilled manual or white-collar workers. Within any segment of the labour market, however, there is a further tendency for disadvantaged groups (such as black workers) to show more variation in unemployment rates than other groups, because employers will tend to discriminate more where they have more applicants to select from.

Out findings thus suggest that, while much the same groups are liable to experience labour-market disadvantages in both growing and declining areas, they would tend to be much better off in the former, particularly in terms of exposure to unemployment. By implication, therefore, either policies to boost employment in declining areas or policies to ease mobility to areas of employment growth would be likely to improve employment prospects for disadvantaged groups from the declining areas. In either case, however, their gains might be substantially at the expense of their counterparts in the areas where employment is currently stable or growing. In other words, the effect of either type of policy might be largely redistributive—between areas, rather than between more and less advantaged groups of workers.

As far as employment stimulation is concerned, our analysis suggests that localized job creation policies in the inner city are unlikely to have much effect on the employment prospects of the disadvantaged unless either the conditions of employment are too unattractive for others to be interested or recruitment is organized in a manner which counters the normal selection criteria of the market. The danger of the former approach is that it may serve to increase the numbers of workers who are effectively trapped in unstable jobs and liable to periodic spells of unemployment—in other words, that it may increase the numbers of those who are actually, rather than potentially, disadvantaged. The latter approach comes much closer to the nub of the problem.

The problem of the diffusion of benefits to other areas and groups of workers arises to a much lesser extent in the case of job preservation as opposed to job creation, while any successes in preventing redundancies would serve to reduce the numbers falling into sub-employment and hence into a disadvantaged status. A stronger case can thus be made for job-saving policies in the inner cities than for untargeted job creation. Our results suggest that job creation policies as such have more relevance to the needs of the declining *regions* of the North and Midlands (or to the national unemployment problem) than to the specific problems of the cities. Throughout this chapter we have noted evidence that at the regional level there are strong links between employment decline and the extent or intensity of labour-market disadvantage, whereas the connection appeared very much weaker within regions. These findings would appear to support the case for conventional forms of regional policy in the northern half of the country.

A number of factors inhibit the movement of disadvantaged groups out of the areas of employment decline. In the case of London, these include the difficulty of obtaining rented accommodation or affordable owner-occupation in the growing parts of the South East and the additional commuting costs imposed on those initially retaining a job in the city by the need to move beyond the Green Belt. Relaxation of either of these constraints would allow these groups a more equal access to areas with better employment opportunities. It could also serve at least to reduce the concentration of disadvantage in the inner areas, and could provide rather better employment prospects for those resident there, as well as mitigating the evident social consequences of high concentration of disadvantage. It would not reduce the scale of labour-market disadvantage on a national scale, however. That would require either a general revival of employment growth—perhaps particularly in the North—or a very determined attempt to redistribute chances in the labour-market. The indications are, however, that any significant redistribution is only likely to be achieved in the context of employment growth and that the persistence of high levels of total unemployment, whether regionally or nationally, seriously inhibits the improvement of the opportunities of those particularly disadvantaged by labour-market processes.

Acknowledgements

This chapter results from research carried out for the ESRC's Inner Cities Research Programme. It draws on data from the 1981 Labour Force Survey, carried out by the Office of Population Censuses and Surveys and supplied to us by the ESRC Data Archive. Responsibility for the analysis and interpretation of these data remains ours alone. We also gratefully acknowledge the helpful comments of Ian Molho on an earlier draft of this chapter.

Notes

1. The measurement of occupational change between 1971 and 1981 requires data published on unchanged boundaries. In practice, because of the effects of local government reorganization this limits comparisons to large urban areas (very few of which have growing employment), 1971 county areas consistent with 1981 counties or groups of districts, and New Towns. The 1981 Census published data for counties, districts, and New Towns based on both the 1970 and 1980 classifications of occupations. The former has been used in Table 4.3 for eight county areas and two groups of New Towns.
2. Outlined by Brown, (1972) and developed in Burridge and Gordon (1981).
3. In this presentation, changes in participation rates are not separately distinguished, being treated in part as 'natural change' and in part as changes in concealed unemployment: see Gordon and Molho (1985) for a more explicit treatment of participation rates. Labour migration is defined to include all residential moves which also involve a change of work-place.
4. As in Burridge and Gordon (1981), net migration is treated as zero in the national equation, with net international migration being subsumed under natural increase, and only differential rates of gain or loss in particular areas being included in the term for net inter-regional migration.
5. Specifically in the form of the migration equation where a *linear* combination of the unemployment rate with a vector of characteristics influencing 'competitiveness' is used to approximate the rate of unemployment attributable to demand deficiency as distinct from structural or frictional factors.
6. Presented in Gordon, (1985b).
7. In fact, the relative position of blacks and the disabled appears to have deteriorated noticeably during this period, while that of less skilled groups of workers has improved somewhat; relative odds of unemployment for the unqualified had not significantly altered; among the young the position was confused by increasing participation in government schemes, but the relative odds appeared to have improved for males and deteriorated for females.
8. This is confirmed by Department of Employment data, at least for Milton Keynes and Peterborough (*DE Gazette*, Feb. 1984, p. 66).
9. That is, white, single persons, in their late 20s, with no disability, an identified sub-degree-level qualification, living in private rented accommodation, in a non-manual or skilled manual job in a stable industry with static employment.
10. One particular limitation is that, as is indicated later, the 25–9 age range exhibits unusually large disparities in unemployment between the three groups of areas.
11. This effect is overstated in the logit estimates since no information was available about qualifications of the over-60s.
12. Structural employment is in this context the aggregate outcome of an increasing supply of labour for unskilled or secondary sector jobs; on the other hand,

increasing *employment* in such jobs, involving above-average risks of spells of unemployment for individual workers, is likely to be associated with higher levels of frictional unemployment.

References

Boddy, M., Lovering, J., and Bassett, K. (1986), *Sunbelt City? A Study of Economic Change in Britain's M4 Growth Corridor*, Oxford: Clarendon Press.

Brown, A. J. (1972), *The Framework of Regional Economics in the United Kingdom*, Cambridge: Cambridge University Press.

Buck, N., Gordon, I., and Young, K. (1986), *The London Employment Problem*, Oxford: Clarendon Press.

Burridge, P. and Gordon, I. R. (1981), 'Unemployment in the British Metropolitan Labour Areas', *Oxford Economic Papers*, 33, 274–97.

Cousins, M. J. and Curran, M. (1982), 'Patterns of Disadvantage in a City Labour Market', in G. Day (ed.), *Diversity and Decomposition in the Labour Market*, Aldershot: Gower, 49–65.

Dixon, R. J. and Thirlwall, A. P. (1975), *Regional Growth and Unemployment in the United Kingdom*, London: Macmillan.

Donnison, D. (1980), *The Good City: A Study of Urban Development and Policy in Britain*, London: Heinemann.

Fothergill, S. and Gudgin, G. (1982), *Unequal Growth, Urban and Regional Employment Change in the UK*, London: Heinemann.

Gordon, A. (1984), 'The Importance of Educational Qualifications to Employers in the Selection of School Leaver Recruits', *Educational Studies*, 10, 93–102.

Gordon, I. R. (1981), 'Social Class Variations in the Effects of Decentralisation in the London Metropolitan Region', paper presented to the *Third International Workshop on Strategic Planning*, University of Dortmund.

Gordon, I. R. (1985a), 'The Cyclical Sensitivity of Regional Employment and Unemployment Differentials', *Regional Studies*, 19, 95–110.

Gordon, I. R. (1985b), 'Unemployment in London', ESRC Inner Cities Research Programme, *London Project Working Paper*, Urban and Regional Studies Unit, University of Kent at Canterbury.

Gordon, I. R. and Molho, I. I. (1985), 'Women in the Labour Markets of the London Region: A Model of Dependence and Constraint', *Urban Studies*, 22, 367–86.

Gordon, I. R. and Molho, I. I. (forthcoming), 'A Model of Occupational Mobility Among Male Workers', Metropolitan Labour Markets Project, Urban and Regional Studies Unit, University of Kent at Canterbury.

Greaves, K. (1984), 'The Youth Opportunities Programme in Contrasting Local Areas', *Employment Gazette*, 92, 255–9.

Hall, P. G., Gracey, H., Drewett, R., and Thomas, R. (1973), *The Containment of Urban England*, London: PEP and George Allen and Unwin.

McCormick, B. (1983), 'Housing and Unemployment in Great Britain', *Oxford Economic Papers*, 35, 283–305.

Norris, G. M. (1978), 'Unemployment, Subemployment and Personal Characteristics: (B) Job separation and work histories: the alternative approach', *Sociological Review*, 25, 327–47.

Owen, D. W., Gillespie, A. E., and Coombes, M. G. (1984), 'Job Shortfalls' in British Local Labour Market Areas: A classification of labour supply and demand trends 1971–1981', *Regional Studies*, 18, 469–88.

Spence, N. A., Gillespie, A., Goddard, J., Kennet, S., Pinch, S., and Williams, A. M. (1982), *British Cities: An Analysis of Urban Change*, Oxford: Pergamon.

5

Local Employment and Training Initiatives in the National Manpower Policy Context

Andrew A. McArthur and Alan McGregor

Introduction

Over the last decade, British manpower policy has been transformed. This transformation has taken place at both the national and the local level. The range of national policy instruments is now much greater and more varied than it was ten years ago. Developments at the urban level have brought local authorities, voluntary groups, and other local interests into the manpower arena where formerly they had little or no presence. The national and the local developments were born out of the same problem—the declining competitive position of the British economy and the consequent dramatic rise in unemployment. Within a short space of time, locally organized manpower, training, and related measures have mushroomed throughout Britain and Europe (see, for example, EEC 1983).

At the national level, manpower programmes have become a much more significant component of government policy over the last decade. The Manpower Services Commission (MSC) is now a high-profile organization with programmes affecting hundreds of thousands of people in any given year. The range of programmes has developed considerably with moves into self-employment in the 1980s—through enterprise training and the Enterprise Allowance Scheme (EAS)—and the development of an extended role in the youth sector with the establishment of the Youth Training Scheme (YTS) in 1983. Most recently, adult training has been revamped with the launching in 1985 of the Adult Training Strategy (ATS).

This growth can be seen both as a response to rising unemployment and as a long-run strategic change in the emphasis given to certain manpower issues. The treatment of young people illustrates this well. The schemes developed from the mid-1970s which were consolidated under the Youth Opportunities Programme (YOP) in 1978 were very much directed at the problem of rising youth unemployment. However, these were essentially presented as 'mopping up' exercises, as evidenced by the temporary nature of YOP. The YTS, on the other hand, is a permanent scheme which can be clearly located within the context of a wider strategy aimed at reforming Britain's system of training (MSC 1981).

The dramatic growth of local initiatives can be explained in a number of ways. First, the collapse of employment levels, which was particularly severe in the period 1979–82, led to dramatic increases in unemployment in urban areas nation-wide. The failure of central government to significantly reduce the jobless level, together with the uneven spatial and social distribution of unemployment, has precipitated local responses to the problem. A second factor has been the failure of national policy initiatives to keep pace with the growing unemployment problem. Although manpower programmes have expanded significantly over the last ten years, the absolute gap between need and provision has widened, especially for those disadvantaged groups such as the long-term unemployed. This growing gap has created space for the development of local initiatives. Third, pessimism abounds about the recovery of the economy along conventional lines. Few national forecasting agencies envisage any significant decline in unemployment. Indeed, although employment levels have recovered to some degree since 1983, unemployment has continued to rise. This pessimism has stimulated unconventional approaches to local economic initiatives, such as community businesses, which have sought to create employment for the long-term unemployed on a not-for-profit basis.

In this chapter we report on research carried out as part of the Economic and Social Research Council's Inner Cities Research Programme. The objective of the research was to consider the relevance of the changing mix of national manpower programmes for the problems of urban areas, and to examine the actual and potential role of more locally based initiatives. The pursuit of this latter objective constituted the primary orientation of the research, with effort directed towards identifying constraints on the effective operation and innovative development of local manpower programmes.

The analysis of national programmes was based on published and unpublished reports and documents, interviews with key national officials (principally of the MSC) and feedback from those involved in locally based initiatives. In the absence of a comprehensive sampling frame, the local initiatives to be investigated were chosen using a variety of sources: evaluative studies, consultancy reports, newspaper cuttings, and, most importantly, word-of-mouth contacts with local authorities, voluntary sector groups, central government departments, and so on. Visits were made to or contact established with the organizers of local programmes which appeared to be particularly innovative. A conscious attempt was made to include a range of activities (for example, training for jobs, self-employment schemes, and basic skills training) and to cover a number of different urban centres.

The findings are organized in the following way. The next section of the paper provides a summary statement of the labour-market problems of the conurbations. This provides a yardstick against which to measure the appropriateness of the national and local responses. The policy discussion begins with the national policy framework and the changes that it is undergoing.

The roles of local authorities and non-statutory bodies in manpower policy are then discussed. This leads into the major section of the paper where there is a detailed discussion of individual initiatives and the lessons that can be drawn from their experiences. A number of recommendations are made at the close of this section. The chapter concludes with a brief summing up of the main points.

Labour Market Problems in the Conurbations

To test the appropriateness of any set of policy initiatives and responses it is important to begin with a clear statement of the problem which is being confronted. From existing studies we are able to summarize the nature of the labour-market problems of the conurbations in the following terms.[1]

First, all the major urban labour markets are characterized by massive excess labour supply relative to current and prospective demand by employers. As a result of the collapse of employment between 1979 and 1982 there is no major urban area where unemployment and vacancies are remotely near to balance. In the more depressed conurbations the situation is severe. For example, in Glasgow in 1984 there were around forty unemployed for each vacancy. Therefore, all the major urban areas need a stimulus to employment to gain purchase on their unemployment problems.

Second, although there is evidence of widespread structural change, there is little in the way of structural unemployment, and this tends to be concentrated in the more buoyant labour-market areas. By structural unemployment we mean a situation where some proportion of vacancies cannot be filled from the ranks of the local labour force. Although traditional industries have declined, growth in new sectors has been modest. Furthermore, this growth has been concentrated in service sectors where skill content is not necessarily high. Growing areas like insurance, banking, and finance recruit, in the main, qualified school-leavers who then acquire their skills on the job, with supplementary college-based instruction. In a recent study (MSC 1985) of 'hard to fill' vacancies in occupations requiring skills up to the craft level, only 5000 were identified, and these were spatially concentrated, with 54 per cent of the total in the South East of England. By definition, skill shortages are likely to be a greater problem in the labour markets for higher skilled occupations. Several recent reports on the information technology sector report a serious problem (Department of Trade and Industry 1984, 1985), although quantification is difficult. However, overall it is likely that the number of vacancies which cannot be filled from current stocks of labour is low relative to the aggregate volume of unemployment. This conclusion holds most firmly in the North where labour markets are generally less buoyant.

Third, although structural unemployment is not currently a serious problem in most urban areas, the future position is less clear. Of particular concern here is the dramatic decline in training investments undertaken by employers. This is most visible through the virtual collapse of apprentice training, which was traditionally a major source of training investments, particularly in the industrial conurbations like Birmingham and Glasgow. With an eye to the future, technician training also appears to be inadequate. For example, while it is estimated that 6000 engineering technicians should be trained annually to meet future demands, the actual figure in training is 3000 (Department of Trade and Industry 1985). Therefore, to the extent that current training is tied to current needs, problems will develop in the future if demand presses ahead in sectors with high training requirements.

An important extension of this argument can be made in the case of particularly depressed conurbations. For example, on Merseyside where training investment has historically been low and in the West Midlands where craft apprenticeship training has been savagely cut, the capacity of the local economy to attract new investors, sustain expansion of growth sectors, and create businesses in new industries is likely to be severely debilitated. Training investment is now accepted to be an important facet of *national* economic development. This is clearly recognized in the government's strong emphasis on reforming Britain's training system. The statistical evidence on Britain's relatively poor performance in the field of youth training was published in a review of the Employment and Training Act 1973 (MSC 1980). More recently, a highly influential report by the National Economic Development Council (NEDC) and the MSC underlines the inadequacies of Britain's system of vocational education and training in relation to the efforts of Germany, Japan, and the USA (NEDC 1984).

Since 1979, major changes in training have been instituted throughout the age range and across all levels of training. These innovations have included the Technical and Vocational Education Initiative (TVEI), the YTS, and the ATS.[2] The essence of government thinking on training and its relationship to the development of the economy is clearly stated in the two 1985 White Papers on employment and training.[3] Referring to the NEDC *Competence and Competition* report, the White Paper on training for young people notes that:

The report underlines the Government's view that vocational education and training are not marginal activities, but are central to our economic growth and prosperity. (HMSO 1985b, para. 2.)

The arguments applied to the link between *national* training investments and economic growth would appear to be equally valid in the context of *local* economic development.[4] This point is particularly relevant given the comparative failure of regional policy to rectify regional imbalances. Traditional regional policy leaned heavily on a theory of regional growth which stressed investment in physical capital as opposed to human capital. Similarly, economic regeneration programmes pursued by local authorities since the mid-

1970s have given much higher priority to traditional measures (for example, small factory construction) than to manpower programmes (Mason 1983).

Fourth, although long-term unemployment is a major national problem, it is particularly severe in the urban areas outside the South of England. For example, the average duration of unemployment in April 1985 was 47 weeks in the West Midlands and 43 weeks in the North West of England versus 30 weeks in the South East of England. Whereas it is clear that the collapse of employment is the fundamental explanation for the rise in long-term unemployment, it cannot be safely assumed that a rise in labour demand will lead to a fall in the numbers who are long-term unemployed. Although there is disagreement on the effects of unemployment on physical health (see, for example, Brenner 1979; Gravelle *et al.* 1981; Forbes and McGregor 1984) there can be little doubt about the impact of longer-duration unemployment on psychological well-being (Warr 1983). Additionally, there are problems of skill loss, both in the specific sense of job-related skill and in the general sense of being able to operate in the workplace environment, accept its disciplines, and so on. As a consequence, investors may stay clear of urban areas where the long-term unemployed bulk large in the labour force.

Fifth, apart from the question of local economic development, many urban areas are characterized by significant inequalities between different groups in the labour market. The changing dimensions of inequality over the period of collapsing labour demand have been examined most exhaustively for London (Buck *et al.* 1986). The London study concludes that labour inequalities have intensified, and demonstrates that in periods where supply greatly exceeds demand, even in relatively buoyant areas like the South East, employers are able to exert an enhanced degree of choice leading to the increased exclusion of disadvantaged groups (such as racial minorities). More generally, as the nature of the unemployment experience has changed, the gap between the employed and the unemployed has widened. Average spells of unemployment now last much longer than they did in the 1970s. The experience of unemployment has changed qualitatively as a consequence. Job generation *per se* will not solve this problem. The longer-term unemployed will tend to be among the last groups to be hired in an upturn of economic activity. Research has shown that new employing units are reluctant to hire from the unemployed. This applies to large mobile investors (Goodman and Samuel 1966) and small entrepreneurs alike (McArthur 1984).

The issues of local economic development and disadvantage in the labour market are, of course, linked over the longer run for any single urban area. These links are of two main types. First, there are the potentially damaging social externalities associated with gross inequalities and the concentration of economic disadvantage among specific groups or communities. Second, as we noted earlier, sustained unemployment may debilitate large sections of the labour force of an urban area. This was a fear expressed in the 1930s by many commentators. The most concise expression of this concern was the Pilgrim Trust's conclusion that ' . . . unemployed men are not simply units of employability who can, through the medium of the dole, be put in cold storage and taken out again immediately they are needed. While they are in cold storage things are likely to happen to them' (Pilgrim Trust 1938, 67). However, mobilization for the war effort meant that the worst fears on the labour-force front were never realized.

This completes the discussion of the labour-market problems of urban areas, particularly those in the more depressed regions where unemployment is high and prospects are bleak for future growth. In the next section, we consider the current provision of national manpower programmes, the trends in programme development, and their appropriateness given the problems identified above.

The National Policy Framework

The growth and change in national manpower policy can be illustrated in part by reference to recent expenditure trends across main MSC programme areas (Table 5.1). A number of features are worth highlighting from the figures in Table 5.1. First, at a time of public expenditure restraint, MSC expenditure has risen very substantially in cash and real terms. Second, the mix of programmes has altered dramatically over a very short space of time. Since 1980, a number of key changes can be identified:

(*a*) Employment services (mainly the Jobcentres) have declined greatly in proportional significance

(*b*) The proportion of the MSC budget spent on adult training has been halved over the six years and has also fallen significantly in real terms

(*c*) Youth training is now the largest budget item and will grow still further with the implementation of the two-year YTS

(*d*) Job creation expenditure grew dramatically over the period of the most rapid increase in unemployment; a further major rise in expenditure was announced in the 1985 Budget.

The general picture is of major change within a very short time period. For example, the EAS, the TVEI, and non-advanced further education (NAFE) were scheduled to take 10 per cent of the MSC budget in 1985–6, starting from scratch in 1983. The extent of change would have been even more vivid if an earlier base year had been chosen. For example, in 1978–9, youth training captured only 10 per cent of the budget and job creation 1 per cent compared to 1985–6 figures of 37 per cent and 26 per cent respectively.

In addition to MSC programmes, the Department of Employment runs a number of subsidy schemes directed at the labour market. These have generally been in decline. Some schemes are in the process of being phased out, while others have already finished. For example, the Temporary Short Time

Table 5.1. *Cash Expenditure (£m.) on Selected MSC Activities*

Activity	1980–1		1983–4		1985–6	
Employment services	132	(15)	144	(8)	145	(6)
Adult training	248	(28)	238	(14)	289	(13)
Youth training	215	(25)	776	(44)	835	(37)
Job creation	50	(6)	399	(23)	592	(26)
EAS	—		23	(1)	115	(5)
TVEI	—		7		39	(2)
NAFE	—		—		69	(3)
Total MSC expenditure	876	(100)	1768	(100)	2262	(100)

Note: The 1985–6 figures are for planned expenditures. The figures for 'Youth training' refer to YOP and the YTS. 'Job creation' refers to the Special Temporary Employment Scheme (STEP) and the Community Programme (CP.) The figures do not include changes introduced in the 1985 Budget which principally affected the CP and YTS. The figures in parentheses are percentages of total MSC expenditure.

Sources: MSC *Annual Reports* 1980–1 and 1983–4. MSC *Corporate Plan 1985–9.*

Working Compensation Scheme (TSTWCS) ended in 1984 and the Young Workers Scheme finishes in 1986, although there are plans to replace it with a New Workers Scheme. The only other scheme of note is the Job Release Scheme which seeks to promote early retirement from the labour force provided the job vacated is subsequently to be refilled.

Having identified some of the key urban labour-market problems and the main thrust of government manpower provision, we can now present a brief assessment of current and prospective national provision in relation to the nature of the problem. To give this discussion some coherence, programme provision is keyed into two specific labour-market issues:

(1) The lack of employment opportunities
(2) The training needs of various areas.

Distributional questions are also important and are dealt with as they arise. To help focus the discussion, a simple distinction is drawn between the more and the less buoyant labour-market areas.

Employment opportunities

Three broad programmes apply in this area: job creation programmes, of which the Community Programme (CP) is the current version; employment subsidy schemes; and self-employment measures. Each of these provides an alternative to unemployment in either conventional but subsidized jobs, temporary jobs, or self-employment.

A major advantage of direct *job creation* measures is the capacity to target them at specific groups. British job creation measures have traditionally been targeted at the long-term unemployed. Given this orientation, the depressed conurbations capture a disproportionate share of CP places, as Table 5.2

Table 5.2. *Community Programme Filled Places and Size of Labour Force by Region*

Region	CP places		Labour Force %
	No.	%	
South East	22 418	17	35
South West	8 491	6	8
Midlands	25 568	19	17
Wales	9 473	7	5
North West	21 307	16	12
North	13 184	10	6
Yorks. and Humberside	14 284	11	9
Scotland	18 060	14	9
Total	132 785	100	101

Sources: MSC and *Employment Gazette*, Oct. 1984.

clearly shows. For example, whereas the South East houses approximately 35 per cent of Britain's labour force, it gained only 17 per cent of total CP places. On the other hand, Scotland had 14 per cent of CP places although its labour force was less than 10 per cent of the national total.

As Table 5.1 shows, job creation places have expanded greatly over the last few years. The projected number of CP places was expanded further from 130 000 to 230 000 at the 1985 Budget.[5] However, there is still a significant gap between need and provision. Given the seemingly inexorable rise in the numbers of the long-term unemployed, even allowing for expansion of the CP to 230 000 places, there are still seven long-term unemployed for each available place.

Various criticisms have been levelled against the CP and its predecessors. A number of these are taken up in more detail later in this chapter where local initiatives are discussed; here we note two general problems. First, a recent evaluation found that 51 per cent of former CP participants were unemployed around eight months after leaving the scheme (MSC 1984b). Although these figures are not necessarily evidence of failure given the disadvantaged client group catered for, the criticism has been made that the CP does not raise employability sufficiently because the jobs created are generally unskilled. The main job areas are construction, horticulture, clerical, painting, transport, and materials handling (MSC 1984a). The government responded to these criticisms by setting aside funds in 1985 for 50 000 CP training places. However, provision has been made for very basic training, to be done in the participants' own time and without benefit of allowances other than for travel or subsistence.

A second potential problem is the ability of different urban areas to meet the expansion targets for the CP. Places under the programme are delivered largely by local authorities and the voluntary sector. The ability of local authorities to expand the number of places on offer will depend in part on the co-operation of local trade unions in that sector. It is likely that a significant proportion of the expansion may fall upon the voluntary sector. Whereas some urban areas traditionally have a very active voluntary sector (for example, Merseyside), others are more dependent on local authorities to supply places. The ability to continue targeting the CP on the areas of highest unemployment may be constrained by the variations in delivery capacities between areas. We return to this issue in a later section.

Employment subsidies, like job creation schemes, have been targeted at specific groups. One of the main recent subsidy schemes, the TSTWCS, was targeted on workers threatened with redundancy. As a consequence, the subsidy tended to go disproportionately to those regions whose industrial base was most under threat. Although employment subsidy schemes have fallen into decline at the national level, as we shall see later a number of local authorities have implemented them in their areas. Some of the difficulties with the effectiveness of employment subsidy measures are discussed in a later section where local authority activities are dealt with.

As national subsidies to employers to hire more labour have declined, a natural replacement has come forward in the form of a subsidy to *self-employment*—the EAS, introduced on a national basis in 1983. The EAS offers a £40 per week allowance for 52 weeks to certain categories of the unemployed setting up in business on their own account. As Table 5.1 shows, the EAS has developed rapidly and in 1985–6 accounted for 5 per cent of MSC expenditure, with over 60 000 starts per annum.

At one level the EAS can be seen as a measure to mop up unemployment by transferring people from the ranks of the unemployed to the self-employed. To the extent that the move is permanent for a good proportion of participants, the EAS may make a small contribution to the longer-run unemployment problem of an area. The evaluations carried out to date (Department of Employment 1984) suggest that large proportions of EAS participants remain in business after the allowance ends. A survey of participants who took part in the pilot phase of the EAS found that, of the 70 per cent who responded to a questionnaire survey six months after the exhaustion of the allowance, 80 per cent were still trading.

A proper assessment of the scheme needs also to take account of deadweight (that is, people receiving the allowance who would have set up in business anyway) and displacement (that is, the extent to which existing output and employment is damaged by competing EAS activity). The evaluation of the pilot EAS estimated deadweight at around 50 per cent, but no evidence on displacement, which is extremely difficult to measure, was collected. The attractiveness of the EAS to government is its apparent high cost-effectiveness and the lack of serious administrative problems. For the pilot EAS the net cost[6] to the Exchequer per person off the unemployment register in the first year was £2690. However, if one assumed that about 60 per cent of the firms that survived the duration of the scheme continued to trade until the end of their second year, then over the two years the net cost of removing one person from the register fell to £650 (Department of Employment 1984).

Although the EAS appears to be a successful national programme, there are problems with the spatial distribution of participation. For example, whereas Scotland has 11 per cent of Britain's unemployed, it has had only 8 per cent of EAS participants. Conversely, the North West of England has 14 per cent of the unemployed but 20 per cent of all EAS participants. There appears to be a relationship generally between a region's tradition of self-employment and EAS participation. As the scheme is essentially demand led, it is tending to reinforce existing patterns. A similar result holds within urban areas if Glasgow is at all representative, with peripheral housing estates having particularly low EAS take-up (McArthur and McGregor 1986).

In general, with the growth of the CP and EAS, the MSC's expenditure has increasingly reflected a need to fill the gap between labour supply and demand. However, this gap is now enormous, and programmes such as these can only touch a minority of the unemployed. The concern for certain urban areas is their capacity to deliver sponsors (for CP places) and participants (for EAS allowances). As the national schemes for mopping up unemployment can reach only a minority of the unemployed, and given the possibility of increasingly variable uptake across urban areas, the scope for local initiatives is large.

Training opportunities

The aggregated figures in Table 5.1 show clearly a significant change in the MSC's training effort. Adult training has declined in significance while youth training has grown.[7] The projected growth in TVEI and NAFE expenditure will accentuate this trend.[8] These changes are significant enough, but they also mask important developments taking place within some of the main expenditure headings. To progress the discussion on these developments we consider youth and adult training separately.

A major development in *youth training* was the introduction of the YTS in 1983. This can be viewed as a natural extension and enhancement of the old YOP scheme which had been widely criticized, mainly because of the exploitation of youngsters as cheap labour with knock-on effects for the unemployment of unskilled adults. On the strategic front, the YTS can be seen as an attempt to introduce a comprehensive, low-cost training system based on the West German apprenticeship system. Our concern is not to mount a critical evaluation of the YTS as a national programme but rather to consider the implications of the YTS and its prospective future development for urban areas. A difficulty here is that the YTS is a relatively new programme which is still undergoing change. For example, in 1985 the government announced that the YTS would become a two-year programme as from 1986.

As the YTS is funded to provide comprehensive coverage to all unemployed minimum-age school-leavers, it is not immediately apparent that its impact will be differentiated spatially. However, there is a significant qualitative difference between regions in the type of provision which is available. When the YTS was established, a structure based on different modes of provision was adopted. Mode A provision was to be employer-based and Mode B a mix of local authority, voluntary sector, and similar provision. Mode A provision was somewhat cheaper, in part reflecting the needs of the more disadvantaged client group likely to be disproportionately located in Mode B schemes.

Regional differentiation has emerged in the distribution of YTS entrants by mode, as Table 5.3 clearly demonstrates. Although the relationship is not entirely clear-cut, there is a strong tendency for Mode A provision to be significantly above the national average in the more buoyant labour markets of southern England and below the national average in most of the northern areas. Indeed, the situation is more complex than this. The Midlands appear to do rather well with 76 per cent Mode A provision. However, in the West Midlands 56 per cent of the Mode A provision is not provided by employers but by private training organizations which have sprung up with the development of the YTS. In smaller areas these organizations completely dominate the Mode A provision.[9]

Table 5.3. *Percentage Distribution of YTS Entrants by Mode, 1985*

Region	Mode A	Mode B1	Mode B2	Total
London	75	19	6	100
South East	81	16	3	100
South West	80	18	2	100
Yorks. and Humberside	70	22	8	100
Midlands	76	22	2	100
Wales	67	29	4	100
North	60	30	10	100
North West	68	27	5	100
Scotland	75	22	3	100
Britain	72	23	5	100

Source: MSC.

The government wishes to see Mode A provision expand because it is cheaper, but also because of a belief that training at the work-place is more meaningful for the youngsters involved. It has been made clear that the proposed expansion of the YTS from one to two years is to be financed largely by employers. Additionally, the division into modes is to be replaced by a distinction between basic and premium funded places. Premium places are intended to cater for the more disadvantaged youngsters; but they are also meant to underpin provision in areas where a sufficient number of employer-based places is not forthcoming. However, taking away the labels will not help to reduce the problem experienced by areas like Merseyside in generating high numbers of good-quality employer-based schemes. This difficulty is directly founded on the decline in the employment base of a number of our larger urban areas.

The differential experience between regions may have an important bearing on the relationship between training and local economic development. If employer-based training is superior over a long period then, other things being equal, the economic gap between the more and less buoyant labour-market areas will widen. The early evidence from the YTS is that youngsters leaving employer-based schemes have better chances of finding a job than those in Mode B schemes. Around 66 per cent of Mode A leavers are finding work compared to 49 per cent for Mode B2 and 33 per cent for Mode B1 leavers. However, the employer-based schemes contained a disproportionate number of the better-qualified youngsters, and this may explain their higher placement rates. Additionally, being with an employer *per se* may raise placement rates simply because a proportion of the young people may well be hired by the sponsoring employer anyway. This may have nothing to do with training but rather the advantages of the scheme as a screening mechanism for employers wanting to hire youngsters. In these cases, successful youngsters will have demonstrated their reliability and good work habits during their spells of work experience.

However, if employer-based provision is superior, equality of training opportunity across regions and urban areas would require an additional resourcing element for those areas with a low capacity to generate employer-based schemes in the required volume. There is some recognition of this in the new arrangements for premium funding. However, to the extent that it is felt to be inadequate, scope exists for local authorities and other relevant local agencies to make good the shortfall in training quantity and quality. This might take a number of forms including the substitution of off-the-job training for employer-based experience and the promotion of joint public–private sector initiatives to generate more employer-based placements.

The field of *adult training* has also been in a state of flux and under pressure from a number of influences. Traditionally, MSC's main adult training programme was the Training Opportunities Scheme (TOPS) which was directed at the unemployed. The great bulk of TOPS courses were for occupational training in skills which it was hoped would lead directly to jobs. The second major element was work preparation courses which were more basic than occupational training. These could lead on to jobs or to further, higher-quality training. The third element which was developed during the 1980s was training for enterprise. Well over 50 per cent of TOPS training was provided in colleges, just over a third in Skillcentres, and most of the remainder in employers' establishments.

Developments in the policy area have dictated a greater concentration on the economic as opposed to the social role of adult training (MSC 1983;

HMSO 1984). Increased emphasis was given, under the ATS, to meeting the skill needs of employers. This change of emphasis reflects a belief in the key strategic role of training in the process of economic development. Additionally, as a result of the collapse of labour demand over the period 1979–82, the placement of TOPS trainees in jobs fell to relatively low levels. Placement rates rose again from around 1983, but this was only after a major cut-back in courses with poor placing rates. Between 1980–1 and 1983–4 the occupational training places available for the unemployed under TOPS fell from 60 000 to 50 000. Over the same period, unemployment rose substantially. The numbers completing the lower-level work preparation courses rose from 6000 to 14 000, and the number completing enterprise training from 400 to 2500.

With the full implementation of the ATS from 1985, these trends will be accentuated. The following are the main elements:

(*a*)   An expansion of enterprise training
(*b*)   The institution of local grants to employers
(*c*)   Increased use of work preparation and of fees-only training (that is, part-time training where the trainee subsists on unemployment or supplementary benefit and does not receive a training allowance).

The latter development will allow the MSC to increase significantly the numbers of unemployed trained at some level. However, as the adult training budget will not rise, the average level of training given must fall.[10]

There are potentially serious implications for those urban areas where the labour market is most depressed. Simply, to the extent that the adult training system is to become increasingly employer-led, resources will flow disproportionately to those areas where employer needs are greatest. These are areas which can most readily demonstrate skill shortages. As we noted earlier, the majority of hard-to-fill vacancies are in the South East. On the side of economic development, the gap in training investment between the high- and low-unemployment areas will grow. Training capacity has already been cut back. The closure of twenty-nine Skillcentres was announced in January 1985. These were disproportionately located in the more depressed areas,[11] where less buoyant labour markets lead to low placement rates.

From the perspective of the unemployed living in more depressed urban areas like Merseyside or the West Midlands, the probability of finding an occupational training place will decline from its already low level. This may be offset by an increase in the relative provision of low-level work preparation courses representing an expansion of existing courses together with the development of CP-linked training. Alternatively, the increase of fees-only provision may permit the resources going into training courses to be maintained by savings on allowances. In these circumstances the trainees bear the costs of the cuts through the difference between their benefit incomes and the training allowance they would otherwise have received. Whichever scenario is correct, the prospects are not good for the high-unemployment areas. In these circumstances there is a genuine need for local initiatives directed at both the developmental and distributional problems referred to above. If areas like Merseyside and the West Midlands do not look after themselves in respect of their training interests, they may slip even further behind in the competitive struggle between urban areas for tomorrow's jobs.

This completes the discussion of national manpower programmes. However, the involvement of other organizations in the manpower area has increased both through sponsorship of key MSC programmes and the independent initiation of local projects in their own right. In the following sections we look first at the role of public agencies, principally local authorities, and secondly at the activities of the non-statutory sector.

Local Authority Activity

Since the early 1980s, important changes have taken place in the approaches to economic development adopted by local authorities. Local action has broadened from a traditional concern with land assembly, industrial estate development, and industrial promotion to include, among other things, manpower and training initiatives. Growing local authority involvement in this area stems from a number of related influences. There has been a gradual recognition that industrial infrastructure investment, such as extensive small factory construction, has had many weaknesses in terms of redirecting opportunities towards residents in the more disadvantaged urban communities. For example, research into the employment effects of such programmes in the Glasgow conurbation found that very few of the jobs generated went to locally resident, long-term unemployed workers (McArthur 1984). Additionally, the effectiveness of promotional campaigns designed to attract footloose industry has also fallen. As a result of the recession there has been less mobile investment around, and competition for it among local authorities has intensified (Middleton 1981).

More specifically, the spur to local authority involvement in manpower measures has come from changes in the structure of local industry and the direction of national manpower policies. The implications of recession have meant, for many areas, a massive reduction in the local training investment and capacity as key employers have contracted or closed. Clearly this must concern the local authorities involved. The employment prospects of residents will depend in large part on their ability to secure training and develop skills. Similarly, the prospects facing the local economy will be influenced by the quality and quantity of its human capital stock.

In addition, as we noted in the earlier discussion, the government has forced through major changes in Britain's system of industrial training through the work of the MSC. On one hand, the expansion of MSC special programmes such as the CP and YTS has facilitated the involvement of local authorities as sponsors of these programmes. On the other, however, the strong employer orientation of training policy has concerned a number of local authorities, particularly those in areas which have experienced significant structural decline. Perceived weaknesses or shortcomings within government manpower policy may well stimulate local authorities to look for alternative approaches to local training which adopt a longer-term strategic approach that is better geared to the needs of the local economy and the problems faced by disadvantaged workers in the labour market. Indeed some have begun to move in this direction.

Some of the above sentiments are clearly embodied in the following extracts from a report by the Greater London Council (GLC):

The collapse of quality training in Greater London in recent years poses real and serious dangers to our city. Without the skilled men and women to support the public and private sectors of the economy, we cannot hope to prosper . . . At a time when private firms are cutting back on apprenticeships and Government is destroying large parts of Britain's training infrastructure, the need for major interventions by local authorities is growing month by month. (GLC 1984, 4.)

While many local authorities are now involved in manpower programmes in some way, the nature of their involvement varies widely. Most have some role in the delivery of the YTS and CP. The careers services of those authorities with education powers frequently act as managing agents for the YTS and control a series of sponsors such as Information Technology Centres (ITeCs), Training Workshops, and Community Project Agencies. The YTS is also internalized, some authorities using it as a form of replacement for the former apprenticeship provision, with trainees unionized and offered permanent jobs when vacancies arise. Local authorities also act as managing agents for the CP, providing nearly half the places available nationally.

An important determinant of the degree of local authority participation in MSC schemes is the attitude of local politicians and trade unions. The extent of involvement has been restricted in some cases because of fears of job substitution, concern about the quality of the schemes, and mistrust over their political motives. However, even in cases where local interests are critical of the YTS and CP, local authorities have sought to use and build upon them. The supply of funds to top up the allowances or wages that individuals receive under the YTS and CP and the provision of grants to enable MSC schemes to buy equipment and thereby make better quality training possible have become quite common. Leeds City Council, for example by spending £8 million on top of £5 million from the MSC under the CP was able to recruit 1300 long-term unemployed and increase the £60 average available under the CP to provide full-time work at union rates. Other authorities have attempted to develop training programmes to complement the CP.

An important contraint on the utilization of CP labour is the limited range of activities in which these workers can get involved. In terms of house renovation, for example, local authority workers funded under the CP can undertake specific tasks such as painting, decorating, and roof repair (under a certain height). Their involvement in electrical work, plumbing, heating, and ventilating may well excite, for good reason, objections from architects' departments and other direct labour workers. Constraints such as these can make it administratively difficult for authorities to put together a package for housing renovation which involves both CP and other local authority labour.

A large amount of local authority activity in the manpower and training area is independent of the MSC and its programmes. A small number of authorities have sought to adopt a strategic approach to training as part of their approach to local economic development. The largest effort of this sort to date has been by the GLC. Highly critical of current manpower policy, the GLC has established the Greater London Training Board (GLTB) as a council committee which draws together a wide range of interests including representatives from the Inner London Education Authority, women's groups, ethnic minorities, trade unions, and the business sector. It is charged with several responsibilities which include: overseeing of the production of a Labour Plan for Greater London; the provision of, and financial support for, training programmes and opportunities to meet deficiencies in the local facilities available; the development of the interface between training and work, and a wider understanding of the workings of the local labour market; and the promotion of debate and awareness about the role of training and local labour-market issues. A further aim is to link the board's programme with the GLC's broader industrial strategy.

In 1984–5, the board spent most of its £7.25 million budget on some fifty projects providing 2600 training places which together have helped to create over 200 jobs in training and related activities. Almost 70 per cent of the GLTB's expenditure went on adult training. This contrasts with the MSC, half of whose London expenditure in 1984–5 was accounted for by YTS allowances. Almost 22 per cent of the GLTB's budget went on enhanced youth and apprenticeship training, with the remaining 7 per cent directed towards research and training support and measures to increase take-up of training opportunities (GLC 1985a). Much of the GLTB's effort has been directed towards designing high-quality training initiatives targeted towards workers often subjected to discrimination such as women and the disabled, preserving and enhancing existing training facilities threatened by recession or faced with closure by the MSC, and involving a wide range of community interests in the formulation of policy.

Although other authorities have adopted a similar strategic approach (for example, Lancashire County Council, the West Midlands County Council, and Sheffield City Council), practical responses by local authorities have been mainly concerned with supporting one-off schemes and initiatives. Recruitment schemes, administered directly by local authorities which provide grants to employers who hire, and in some cases train, certain types of workers, are now common. Training projects operating independently of, but supported by, local authorities are also widespread. Voluntary sector organizations can play an important role in the management of these initiatives. A recent study of joint local authority and voluntary sector action found that four out of every five authorities contacted had some involvement with the voluntary sector in employment-related initiatives, although the nature of this involvement was subject to wide variation (NCVO 1984).

In a climate of tightening public expenditure restraint and growing fiscal pressures on local government, the European Social Fund (ESF), which can provide up to 50 per cent of the cost of approved training projects, has been a valuable source of funds for supporting independent local manpower initiatives. Of the £350 million that Britain secured in 1984, the public organizations outside the MSC—mainly local authorities—received one-third. Many of the innovative approaches to local training which exist are underpinned by matching ESF funds. Some examples are cited later in the chapter.

The ESF, however, is not without its problems for local authorities. The major difficulty stems from the fund's operation on a calendar-year budget and the timing of decisions on applications. For projects starting in January of each year, local authority submissions have to be submitted to the Department of Employment in August of the previous year. The European Economic Community (EEC) usually decides on applications the following March, and finance is not forthcoming until some time later. Indeed, in 1985, because of difficulties in agreeing a budget, the whole process was set back even further. The implications of the ESF budget process are that local authorities have to commit funds to cover the full cost of all their schemes not knowing whether these projects will receive ESF support. It may also prove more difficult to secure ESF support in future. In 1985 the fund was heavily oversubscribed as other countries increased their bids for resources. It is therefore unlikely that Britain will attract the proportion of the ESF that it has in the past. In 1984, Britain's share of the total ESF budget amounted to 32 per cent. Without a substantial increase in ESF resources the relative importance of the fund in supporting innovative local manpower initiatives may well decline.

Even with support from the ESF, the resources most local authorities devote to manpower and training initiatives are relatively small compared to those released locally by the MSC. In London, for example, where there is a strong commitment to local manpower initiatives, the GLTB's 1984 budget of £7.25 million compares with MSC expenditure of some £75 million in the same year. This creates a classic dilemma: do authorities spread these resources thinly and go for a wide coverage, or channel them into projects which impact on only a small number of people? West Midlands County Council have developed an interesting approach here by setting up projects which might have a strong demonstration effect. A good example here would be projects trying to get women to train for occupations normally employing male labour. These kinds of projects may have wider spin-offs in terms of their impacts on women's perceptions of employment opportunities and the attitudes of educational institutions, trade unions, and employers.

Apart from the financial constraints on local authorities, those wishing to engage in manpower and training programmes are faced with legal obstacles arising from the lack of statutory responsibility in this field. Furthermore, while government has encouraged a large amount of local authority activity in the economic development sphere—for example, by introducing the Inner-Urban Areas Act in 1978—no similar encouragement has been forthcoming in the area of labour-market policy. Hence, authorities wishing to develop a strong role for themselves in this area must look for loopholes in the existing legislation through which they can legally account for the expenditure they incur. The most systematic attempt to exploit the legislative machinery of local government to this end has been made by the GLC through the work of its GLTB. At the outset, the GLTB identified two main powers which it intended to use: Section 137 of the Local Government Act 1972, which enables authorities to incur expenditure up to the value of a 2p rate product in any one year in the interests of their area, and Section 45 of the Miscellaneous Provisions Act 1982, which allows authorities to use expenditure powers in the Employment and Training Act 1973 to assist MSC schemes. Initially the GLC had intended to use Section 45 as the principal method of accounting for training expenditure, by interpreting the legislation to mean that the authority could do anything the MSC could do, subject to 'agreement' being reached with the commission. Although early developments seemed promising, progress came to a halt following a decision by the government that any non-MSC expenditure subject to a Section 45 agreement would be deducted from the MSC's total training budget even in cases where the MSC was not involved in any actual expenditure. The impact of this decision has been to limit the amount of resources local authorities can spend on training.

For the GLC in particular, the collapse of the Section 45 option has meant that they have had to look for other legal channels. Most of the GLTB's resources have been raised under Section 137, which in London and in many other areas is a popular source of revenue for industrial and employment measures. However, the pressure on Section 137 locally has forced the GLC to look further afield. Section 142 of the Local Government Act 1972 provides powers relating to the provision of information about local services to the local community. These have been used to finance the provision of information about access to training opportunities and to pay the salaries of advice and outreach workers. Potentially, this is a very valuable power in terms of stimulating the take-up of programmes in areas where this is low. Section 145 of the same Act allows authorities to incur expenditure linked to the development and improvement of the knowledge, understanding, and practice of the arts and crafts which serve the arts. This provision, along with

Section 59 of the London County Council (General Powers) Act 1947, which allows the council to fund a body rendering a public service by way of cultural activities, has been used by the GLC to partially fund a training programme for the free-lance film and television industry.

Section 71 of the London Government Act 1963 allows the council to establish an organization which can undertake investigation and collect information related to the local area, and take steps to make this information or the results of studies publicly available. Section III of the Local Government Act 1972 allows the authority to fund research linked to the discharge of its functions under Section 71. The GLTB has attempted to exploit these powers by part funding an open-access project which serves ethnic minorities and provides a comprehensive training advisory and placement service. Powers under the Health Services and Public Health Act 1968 have also been used to fund training-related provisions. Section 65 of the Act has been used to meet the costs of child-care provision to complement women's projects, and to provide training for the disabled. Under Section 65, an authority can provide grants to voluntary bodies involved in providing, promoting, or advertising services to these groups.

The experience of the GLC suggests that, even if an authority commits substantial resources towards a training strategy which attempts to meet local needs which are currently not met or are inadequately catered for by the MSC, the lack of statutory powers can constrain the implementation of the programme. Authorities with education responsibilities which the GLC does not have, such as the Scottish regions and the English county councils, have other powers through which they can provide finance for vocational guidance, training, and other educational programmes. Should the political will to develop a strong local manpower strategy exist in such cases, these authorities may have fewer legal constraints in carrying through a programme.

The Role of Non-statutory Organizations

The involvement of non-statutory organizations in economic and employment initiatives in large urban centres has also grown in recent years. Organizations such as the National Council of Voluntary Organizations (NCVO), the Unemployment Alliance, and the British Unemployment Research Network operate at a national level. Together they fulfil a variety of functions including information and advisory services, forums for debate and pressure group activity, and help for local projects involved in employment and training initiatives. Other large non-statutory organizations are actively involved in establishing local projects. For example, the Community Projects Foundation establishes, jointly with a number of local authorities, community development initiatives, many of which are oriented towards work and training. Project Fullemploy, perhaps the best example of private-sector-led activity in this area, draws on both public and private support and is associated with a series of training initiatives targeted mainly towards disadvantaged ethnic minorities in a number of urban centres. Yet another group, the Neighbourhood Energy Action Project, has been establishing schemes which use labour funded under the CP to form teams of local workers who carry out energy-saving work for poor households in deprived local communities.

Indeed voluntary groups have for some time played a key role in the delivery of MSC special programmes. For example, they have been accredited with setting up some 44 per cent and 25 per cent of former YOP Community Projects and YOP Training Workshops respectively (NCVO 1984), and the NCVO estimates that, in 1985, 50–55 per cent of CP places were provided by the voluntary sector. Indeed, through its participation in MSC programmes, the voluntary sector has been able to exert a modest influence on MSC policy. A recent example of this seems to have been the introduction of a part-time training element to the CP which the NCVO claims was introduced largely as a result of critical feedback from voluntary sector sponsors of CP schemes.

At a local level, the participation of the voluntary sector in manpower and training initiatives can be subject to wide spatial differentiation. In addition to the level of development of the local voluntary sector, this pattern also appears to be determined by the availability of public resources and the encouragement voluntary groups have received from the local public sector to get involved. In England, some of the most active areas appear to have been those designated as Inner Area Partnership or Programme authorities under the Urban Programme. In these cases the vibrancy of local groups is a reflection of the additional resources available and the requirement on designated authorities to work more closely with voluntary organizations. Urban Programme finance has provided a valuable complement to MSC resources for many projects. MSC funds mainly meet revenue expenses, while the Urban Programme can provide capital for premises and equipment and, in certain cases, revenue for topping up supervisors' salaries and paying the wages of additional staff.

In the Partnership area of Liverpool, for example, a powerful role for voluntary groups in manpower initiatives was built up under the former YOP and Community Enterprise Programme (CEP) Schemes—a role which at the time was strongly encouraged by the MSC and Liverpool City Council. By 1982, compared to voluntary groups in other parts of the country, they were claiming the highest number of MSC places. They were also able to draw around £4 million per annum from Urban Aid and Inner City Partnership monies at this time. Following 1982, however, the MSC substantially reduced its funding of the voluntary sector's activities, and support was also almost totally withdrawn by the city council following a change in political control. Nevertheless, the local voluntary sector, by virtue of its strength in the field and new support from Merseyside County Council, was able to turn to the ESF for resources to continue its activities. In 1983, Merseyside's voluntary sector was receiving proportionally more ESF resources than similar groups in other areas of Britain.

The attitudes of local authorities to voluntary initiatives in manpower policy clearly vary widely. Many authorities have encouraged voluntary groups in the management of one-off projects. However, few have sought to develop policies in close collaboration with individuals and organizations in the local community. The GLC is one of the authorities which has, and the types of arguments employed in favour of involving the voluntary sector in the development of a London-wide training strategy identify very well the potentially distinctive contribution local groups may be able to make. The GLC has argued (GLC 1985b) that ' . . . voluntary organizations are likely to be closer to the perceptions and aspirations of the adult unemployed than are the statutory bodies.' The GLC also points out that the voluntary sector ' . . . has been responsible for much innovation and creativity in training, particularly for disadvantaged groups which have not been well served by the statutory and education systems.'

In 1985, the attractiveness of voluntary sector initiatives was enhanced when the EEC indicated that in non-priority areas, like London, voluntary sector applications for ESF support would be subjected to fewer cut-backs ('weighted reductions') than in cases where the local authority was the applicant. Hence, in these areas the ability of voluntary groups to front applications may prove to be a valuable asset in protecting the flow of ESF resources into the local area.

Looking to the future, in the short-term at least, the voluntary sector will be expected to play an increasingly important role in the delivery of key manpower programmes, principally the MSC's CP. In the Easter Budget of 1985, the government announced a 100 000-place expansion of the CP to take place between June 1985 and May 1986. The MSC expect voluntary organizations to take up a large proportion of this expansion. The commission's intention is to build upon existing initiatives in the areas of energy conservation, environmental improvement, and housing refurbishment through the two-pronged strategy of expanding existing national initiatives and encouraging the extension to other areas of one-off local schemes which are considered successful. Outside of the mainstream CP, the MSC introduced a budget of £10 million in 1985–86 to fund locally devised demonstration projects to be directed towards voluntary organizations who would tap this resource as part of their normal activities without having to adhere to the formal CP administrative mechanism.

The size of voluntary organizations involved in local manpower initiatives may have important implications for the resourcing and effectiveness of their programmes. For example, there have been cases where voluntary schemes waiting on ESF funds have been unable to secure tiding-over resources from the local public sector, which is generally the source of matching funds. In these cases, unless the voluntary organization is sufficiently well resourced to provide this tiding-over money itself, the survival of the project may be threatened. The question of scale can also determine the amount of topping-up activity (such as providing extra training) that groups can undertake. For example, small managing agents of MSC programmes who wish to build upon the schemes they run may find it difficult to muster sufficient resources from their managing agents' fees to do this. The Glasgow Council of Voluntary Service report that CP managing agents with under 100 places find it very difficult to develop in-house training programmes and to invest in hardware, course materials, and so on.

While current developments in national manpower programmes may bring extra resources for voluntary groups, they may also impose certain strains. Difficult ideological dilemmas may have to be confronted as a consequence of participating in programmes dictated by a short-term political agenda. For example, the decision to restrict CP eligibility to benefit-recipients sparked off a series of representations from voluntary bodies on behalf of client groups who were excluded from participation as a result. It also exposed the government to the criticism that it had excluded groups in need in order to achieve a greater reduction in the official unemployment total, and hence a lower per capita cost for the scheme. A more practical dilemma stems from an increased dependence on short-term funding. Reapplication on an annual basis is required for schemes based on CP and ESF funding.

A final area of non-statutory activity concerns the role of employers and corporate organizations. The MSC already relies heavily on private sector participation in the delivery of the YTS. In its expansion of the CP the MSC will also be looking to the possibility of employers playing a greater role. Outside the YTS, the private sector is currently heavily involved in training initiatives directed towards small firms. The main contributions in this respect have been the release of personnel to enterprise trusts. These organizations provide a focus for support and advice to entrepreneurs. The more active ones are also engaged in operating schemes on behalf of public authorities and in participating in the marketing and promotion of these schemes.

Innovations in Local Manpower Policy

We have already discussed the main directions in government manpower policy and the redistributive implications they may have for different urban areas. We now turn to consider the experience of initiatives which have sought to tackle the outstanding problems of the disadvantaged and local economic development in new ways. The initiatives we traced in the course of our study included some which tap into MSC resources and others which operate independently of the commission's funds. In particular, we will be looking to identify both opportunities and constraints which appear to confront creative local effort, and to draw on any evidence which exists concerning the effectiveness of these measures so far. First, we discuss man-power and training measures geared to levering people back into jobs. Secondly, we look at the possibilities surrounding the establishment of new economic activity. Thirdly, we highlight some of the key impediments to local innovation and make a number of recommendations as to how the effectiveness of local manpower policy could be improved.

Manpower and training measures geared to employment

Three broad categories of activity can be identified here: first, training and support provisions directly linked to employment opportunities; second, other training measures aimed at raising general skills and aptitudes; third, measures which attempt to influence the recruitment practices of employers and redistribute employment opportunities towards disadvantaged workers. We will now discuss each of them in turn.

Training for jobs As government-funded training becomes increasingly related to the needs of industry, access to vocational training opportunities is likely to become very selective. People who have fewer existing skills to build on and those who have been out of employment for long periods will find great difficulty in moving back into employment via employer-led training programmes. There is, therefore, a substantial disadvantaged client group for vocational training initiatives to address.

Many of the innovative projects concerned with vocational training attempt to fill what can be termed a 'policy gap'—that is, training of a specific type for certain groups of people who are thought to be inadequately provided for by mainstream government measures. The most comprehensive and strategic attack on the policy gap has been made in London by the GLC. In 1984, for example, at a time when government-supported apprenticeship training was falling rapidly, the GLC identified a high level of skill shortages and wholly or partly financed over 750 apprenticeship places in sectors such as motor vehicle repair and maintenance, hotels and catering, and engineering. A large proportion of this training is targeted specifically at disadvantaged workers. As one of a number of initiatives geared to ethnic minorities, an Apprenticeship Training Scheme in ethnic minority businesses has been established with Urban Programme finance to create employer-based training opportunities for unemployed young people. The scheme started with seventeen employers in 1983. At the end of the first year, 95 per cent of the trainees were retained by the firms and continuing with their training. The scheme was subsequently expanded to include thirty-one firms in September 1984 (GLC 1985a).

In Sheffield, the city council has adopted a strong role in women's training. An Equal Opportunities Section, established as part of the city council's

Employment Department, gives priority to women's employment issues. The council also supports a range of voluntary-sector-led training projects which include intensive six-month practical courses for young women covering the first year of the City and Guilds Plastering Certificate, and a Women's Tools Library which provides skill sharing and practice in the use and maintenance of hand and power tools and general building equipment. Many of the projects targeted at women are supported by the ESF. The ESF gives priority to women and is also able to accommodate child-care components which can be vital complements to training if women with dependent children are to have the opportunity to participate.

The effectiveness of vocational training initiatives targeted at the disadvantaged must be judged largely by their ability to reach members of the client group and reduce their competitive disadvantage in the labour market. From the limited evaluative evidence that is available, some initiatives appear to be quite successful, though it should be borne in mind that most of the evidence available has not come from the results of independent research.

One such initiative is the Charlton Training Centre in London. This is a former MSC Skill Centre saved from closure by a consortium of statutory agencies, the GLTB, and other local groups. The initiative reflects the importance that the GLC has placed on preserving local training facilities. Furthermore, it represents an attempt to introduce an approach to training different from that previously provided by the MSC. In particular, attention has been given to removing traditional forms of discrimination in the training process and developing programmes which are responsive to the range of groups and needs in the community (GLC 1985a). The consortium's approach to training involves a high level of local involvement in the planning and running of courses, community access to the facilities outside of normal training hours, and a major expansion of facilities for disadvantaged groups of unemployed people. Of the first forty people trained, twenty-one were women, thirty-two were from the ethnic minorities, and nine had disabilities (GLC 1985b). The degree of participation by the disadvantaged is ascribed largely to the nature of the centre's operation: for example, intensive outreach work, an extensive process of community consultation, the availability of a nursery, and a network of child-minders.

Evidence from other more established initiatives dealing with disadvantaged workers suggests that some have successfully levered their trainees into employment. The Denham Court Training Centre in London, for example, began in mid-1983 as a facility for training disadvantaged young people with trainee allowances paid under the YTS Mode B. The plan for the centre included two particularly innovative aims: some of the trainees would be resident in the building—hence the need for high staff/trainee ratios—and in its operation the centre would develop structures for democratic participation by trainees. Of the first twenty-four trainees who left the centre, fifteen moved into jobs, and another three into further training (GLC 1985a). However, according to the GLTB, a series of difficulties has been experienced in the centre's relations with the MSC. These include:

(a) Disputes over the type of training offered, the MSC feeling that too narrow a vocational bias was being adopted

(b) Intense MSC scrutiny of staff, pay, and recruitment policy

(c) Disputes over trainee catchment areas

(d) Potential disruption stemming from MSC plans to transfer in YTS trainees from other schemes who had only a short period of their YTS year left to complete

(e) The threat of total withdrawal of MSC support as part of the 1984 cuts in the YTS Mode B.

Although MSC support was eventually cut by only half, the survival of this innovative project has largely depended on finance from the GLTB, which in 1984 injected some £297 000.

Successful placement rates have also been reported by Project Fullemploy as a result of its work with disadvantaged adults, most of which has taken place in the South of England. Within six weeks of course completion, 74 per cent of adult trainees were found to be in work or further education. Almost 90 per cent of the trainees were black, the length of unemployment prior to starting the course averaged 17.3 months, and the drop-out rate was lower than one in ten (Project Fullemploy 1984). Project Fullemploy also uses MSC funds to pay trainee allowances and standard staff costs. In this case, however, public resources are supplemented with inputs in the form of seconded staff and financial contributions from private sector sponsors.

While these appear to be encouraging innovations, several notes of caution need to be aired. The training initiatives described above are operating in relatively prosperous areas of the country. In these areas it should be possible to achieve higher placement rates than in more depressed local economies if an equally disadvantaged client group is catered for. While this will be true in general terms, it may nevertheless still be possible to achieve good results in less buoyant labour markets. A recent pilot course in adult foundation skills run by Project Fullemploy in Birmingham reported that eight out of nine participants were placed in jobs at the end of the course. However, an important question regarding the potential of vocational training for the disadvantaged deals with the degree to which success can be maintained with larger numbers of trainees. As existing courses draw small numbers of people from a very large client group, there is a possibility that a creaming effect is operating—that is, those participating are the better-placed among the disadvantaged. Some of the courses, by imposing certain basic entry tests such as literacy and numeracy, are clearly reinforcing the process of selectivity. Although it seems possible to lever small groups of people into jobs, it may well be extremely difficult, particularly in depressed economies, to replicate success with larger numbers and relatively more difficult clients.

Turning specifically to linking the long-term unemployed with job opportunities in severely depressed communities, a novel training programme has been launched in the West of Scotland by Strathclyde Community Business Limited (SCB Ltd.)—a region-wide development unit promoting the development of community business. With support from the ESF, SCB Ltd. has launched a vocational training programme which trains local unemployed people either in key skills that new community businesses require, or to carry out jobs that existing enterprises are planning to undertake. Over the first three months of the programme (October–December 1984) seventy-two people were trained in seventeen different schemes at an average cost of around £1700 per trainee. Participants had previously been out of work for an average of 18 months and SCB Ltd. report that around 50 per cent of the trainees will move into permanent jobs (SCB Ltd. 1985). At least one community business has benefited already by being able to take on board new work and labour as a consequence of local unemployed people being trained in roofing skills. There appear to be relatively few practical examples such as this which attempt to link the training of disadvantaged unemployed people with the development of new and existing business ventures.

Raising the level of general skills and abilities   Local groups willing to innovate in the training area are faced with a major strategic choice of whether to train the disadvantaged in vocational skills or whether to concentrate on equipping people with general skills and abilities. The argument against the vocational option points out that in recession there are fewer jobs and skill-shortage problems than in a climate of growth. Furthermore, it is extremely difficult to anticipate what the future skill demands of industry are likely to be. At a time of low labour demand and intense competition within the labour market, a strategy of vocational training for disadvantaged workers may be a high-cost, high-risk option.

There is, on the other hand, a strong case to be made in support of a strategy which provides people with general skills. There is also an important remedial function to be performed by countering the corrosive effects of lengthening unemployment on personal capacities and aptitudes. A strategy of maintaining and enhancing the general skills of the disadvantaged therefore serves to narrow competitive gaps in the labour market and makes it easier and cheaper for this group to adjust to new skill demands. For reasons such as these, some local authorities have developed substantial general skills programmes. These serve the dual purpose of investing in human capital in the anticipation that economic recovery will eventually come and of meeting the needs of individuals currently excluded from the labour market.

A further attraction of a general skills approach is its role in improving access to further and higher education for disadvantaged groups, the end result being that they are in a stronger position to compete for jobs. A number of initiatives have been established. One in Hackney, London, aims to access local unqualified black workers into business studies training with the longer-term objective of enabling them to compete for management-level jobs in large central-city commercial enterprises—an employment sector where black people are under-represented. The project has arranged a sequential programme of three training components involving a basic return-to-study course followed by one year in a college of further education which leads to a full degree course at a local polytechnic. Early feedback concerning performance suggests that a relatively small proportion may find their way through to a successful completion of the full programme. Of the first intake of thirty, twenty-two got into the further education phase. Only thirteen of these started on the polytechnic degree, although four others were re-routed to lower educational qualifications.

Another scheme based on the same general principle is operated by the University of Glasgow's Department of Adult and Continuing Education with financial support from Strathclyde Regional Council. The department runs a one-year course which caters for around 100 mature students. No formal entry qualifications are required, and bursaries can be offered to unemployed residents of communities designated by the council as in need of priority treatment. Participants who achieve a certain grade of pass in two subjects are guaranteed a place in any faculty of the university except those of engineering, medicine, and science.

The development of general skills programmes may face a variety of constraints in terms of securing MSC support. As we discussed in the earlier section on national policy developments, the MSC has moved away from training for stock and now gives priority to training to meet the needs of existing industry. Furthermore, if the immediate objective of a training project is to link people into an ongoing educational programme, a stark dilemma arises. MSC funds are for training not education. Although in some cases MSC funds have been secured for such schemes, as in the Hackney example above, in future it will be very difficult to convince the MSC that it should fund employment-oriented schemes which link into further training and are unlikely to lead to jobs for some time. Difficulties may also arise in respect of special access programmes which depend on special arrangements with local polytechnics and universities. For the foreseeable future, higher education institutions will be constrained in their student intakes. Furthermore, there may be resistance to recruiting less qualified students in cases where the postgraduate employment experience of participants has a bearing on resource allocation criteria.

A problem which runs through all efforts to provide people with basic skills concerns the difficulty of reaching the more disadvantaged and less motivated sections of the potential client group. People who know about, or at least know how to find out about, training provision pose little difficulty for organizers of local initiatives so long as the training offered is sufficiently attractive to potential participants. Those less motivated by virtue of disillusionment with traditional educational experiences, long-term unemployment, or physical distance from training facilities will be more difficult to reach. Securing the participation of these groups will depend not only on the type of training but on how it is organized, structured, and delivered.

A number of local projects have been exploring creative ways of opening up training provision to a wider and more disadvantaged client group. One large multifunctional centre in Liverpool—Merseyside Education, Training and Enterprise Limited (METEL)—operates two programmes of interest. The 'Training on Tap' scheme provides training in a wide variety of areas and is heavily used by women, the long-term unemployed, and ex-MSC special programme participants. People have access to training at times which suit their requirements without having to face delay imposed by application and screening processes. The 'Earn and Learn' scheme runs in parallel with Training on Tap but focuses on a different client group—principally males on the MSC's CP. Again, open access to training without qualifications is available on a voluntary day-release basis. The training curriculum is not pre-set. Participants can plan their own programme around their requirements from the range of training activities available. Both of these schemes are greatly strengthened as a consequence of METEL operating a variety of other training programmes. Participants on Training on Tap and Earn and Learn are able to access courses and facilities funded under these different programmes and so benefit from a much wider range of resources than that accounted for by an individual programme.

Other initiatives, run principally by local authority adult education departments, have sought to increase the participation of the disadvantaged by taking training to the people and organizing courses within the community itself. In Strathclyde, one programme supported by the ESF operates on the basis of outreach workers first consulting with individuals and groups within depressed communities in order to identify an interest in a particular type of training. Once identified, short courses are then established, taught by further education staff, and often run out of facilities located as close as possible to where the trainees live. By taking the college to the community in an informal, flexible, and demand-responsive way, the initiative represents a reversal of the traditional, centralized, and formal approach to further education.

At the policy level, there appears to be little encouragement to co-ordinate and cross-fertilize key manpower programmes, as METEL is attempting to do. There is therefore a need to develop outreach and integrated capabilities around key programmes in order to reach more disadvantaged client groups and increase the potential effectiveness of existing policy provisions.

Levering the disadvantaged into jobs As we noted earlier, one of the major problems that has bedevilled recent urban policy has been the failure of area-focused regeneration strategies based upon small factory provision to significantly reduce local unemployment rates. Many small firms setting up in areas of priority treatment are unlikely to hire from the disadvantaged or locally available stock of labour, preferring instead the larger pool of better-quality labour available at an urban or regional level (McArthur 1984). For policy-makers concerned with the problems of small areas and disadvantaged groups in the labour market, the possibility of making the hiring practices of local employers more spatially or group specific is an attractive one. An obvious policy instrument in this respect is the recruitment grant. Since 1980, when Cleveland County Council pioneered the use of the ESF to establish a recruitment scheme, the use of recruitment grants designed to lever disadvantaged workers into jobs has become widespread. Individual schemes differ on specifics, but generally they involve a fixed-term percentage wages subsidy which is paid to the employer on certain conditions (for example, that the person hired has been unemployed for a given duration and is guaranteed employment for a defined period after the termination of the grant). Some schemes also incorporate a condition that the workers receive training. Others pay a higher level of grant for the recruitment of disabled workers.

Two such schemes, both part funded by the ESF, were operating in Strathclyde region in 1985 and provide good examples of the concept. The Strathclyde Regional Council operated an Employment Grants Scheme (EGS) which covered the cost of 30 per cent of gross wages (60 per cent if disabled) for firms hiring unemployed 18–24 year-olds or long-term unemployed adults over 25. This scheme operated region-wide. In the financial year 1985–6 some £4.3 million was allocated to the EGS with a projected coverage of 3750 jobs. Between June 1982 and May 1985 the regional council claims to have supported the creation of around 8200 jobs, with 75 per cent of the jobs still in existence after 18 months of the grant. A variant of this—the Training and Employment Grant Scheme (TEGS)—was run by the Scottish Development Agency (SDA). This scheme, however, concentrated on the depressed east end of Glasgow and also imposed a condition that employers provided a certain amount of training to the new recruit.

The apparent success of recruitment schemes like the EGS in terms of levering unemployed people into lasting jobs must be treated with extreme caution. Without further rigorous enquiry it is impossible to determine what proportion of the workers hired would have been taken on anyway. To test these schemes against their objectives we need to establish what leverage has been exerted at the margins to persuade employers to hire more disadvantaged workers than they would otherwise have done. In situations where firms are looking to hire relatively high-quality labour and have a large pool to choose from, it will be difficult to gain a leverage in favour of the disadvantaged.

In a number of cases, recruitment grants will represent no more than a financial subsidy to the employer. Although this is certainly not the main objective of such schemes, the injection of resources into a firm may lead to greater profit and may contribute marginally to local economic development. If recruitment subsidies are having more impact here, and less on leverage, then policy-makers are faced with a strategic dilemma: either they allocate resources to firms in the hope that a healthier local economy will eventually lead to employment spin-offs for the disadvantaged or they look for alternative programmes which will impact directly on the client group in mind. If it is the case that recruitment schemes are operating essentially as subsidies, then there is the outstanding question as to how their effectiveness compares in business development terms to other economic development incentives, such as grants, loans, premises, and so on. This question, however, needs more comparative policy evaluation than is currently available before it can be answered.

Generating new economic activity

For depressed local economies with little prospect of attracting new external investment, the possibility of stimulating new indigenous economic activity as a source of jobs and incomes is an attractive and important option. However, in many of the more deprived communities this task will be particularly difficult for a variety of reasons. These may include a lack of small business tradition, low incomes and limited ability to raise start-up capital, low demand for local services and products, and physical constraints like the lack of small premises and distance from markets.

A growing political importance has been bestowed on the small firm as a source of employment and the principal motor of local economic regeneration. The promotion of this sector by government and public development agencies through a series of physical, fiscal, and advisory stimulants is well established. A more recent area of policy development concerns training. The importance of self-employment as an alternative to unemployment has now been recognized within MSC policy. In 1984 a Training for Enterprise Programme (TEP) was announced as part of the MSC's new ATS. In addition to funding training programmes for new and existing businesses, the longer-term objectives of this programme are to influence the attitudes of educational institutions towards enterprise and to encourage the development of a better training system in this area. Part of the TVEI and the MSC's Graduate Enterprise Programme, which provides short training courses for pre-graduation students with good self-employment prospects, are examples of recent initiatives in this area.

In 1984–5 the TEP budget totalled £8 million. In 1985–6 it was expanded to £14½ million and was expected to increase further in the financial year 1986–7. In the allocation of these training resources an important shift appears to be taking place. Between 1984 and 1986, for every existing entrepreneur trained, four new businessmen were assisted.[12] According to the MSC, in future proportionally more resources will go to individuals who have a good chance of success and existing businesses which have got over the initial hurdle of starting up. The shift is supported on the grounds that existing entrepreneurs have shown they 'have what it takes': hence there is less chance that resources will be wasted on ventures which may well fail at an early stage. It is also argued, with some justification, that small firm pro- grammes have been biased towards training for new starts, with little help available for existing enterprises. The MSC aims to redress what it sees to be an imbalance in this area. Indeed, most of the TEP pilot projects operating in 1985–6 were targeted at existing small business owners or managers.

The implication of this shift may further constrain what can be achieved among disadvantaged workers and communities. At present the main thrust of MSC-backed training for new business starts is concentrated in short courses which look to recruit people who, after a few weeks of training in business or productive skills, or both, will be able to set up and manage a small firm effectively. Given the nature of these programmes it may well be difficult to find recruits from among the disadvantaged or to fill a training course with participants drawn from a particular community. In Glasgow, for example, start-your-own-business courses have been selective in whom they recruit. As a result they have had to draw people from a very wide area in order to meet course target numbers, which in some cases are only a dozen. As they stand, training programmes for new entrepreneurs have been unable to sustain a localized focus at the level of small communities or to involve large numbers from the long-term unemployed.

If it is the case that a proportion of the potential client group for small firm training is not being reached by existing policy provisions, then we can speculate that, to strike deeper into the more disadvantaged sections of that client group, training and support inputs of a different and possibly more expensive sort will be required. A fundamental constraint on the MSC becoming involved in new forms of training stems from the cost-effective criteria imposed by government on new self-employment programmes. Training programmes must be seen to be providing value for money in a strict and limited economic sense. Programmes targeted at more 'difficult' client groups, and which demand more resources to be effective, will find it harder to gain support.

A number of practical ideas and lessons about how to construct effective policies which stimulate new economic activity among disadvantaged groups have been forthcoming from the experience of innovative local projects oper- ating outside MSC self-employment policy. Some of these projects have managed to use MSC programmes not designed to support small business in the process of establishing new enterprise. The CP, for example, has been used as a means of bringing together a work-force with wages and overheads guaranteed, if only for a temporary period, and, during that period, it has been possible to train and learn to manage that work-force, carry out product development and practical market research, and build up a track record which will assist the project to obtain funds from other sources. The use of the CP as a launch pad for new enterprise in this way can be seen in the efforts of community businesses. Several of these projects have successfully achieved the transition between temporary employment and potentially sustainable economic activity (McArthur 1984). However, CP projects are not generally allowed to retain income generated through commercial operation, nor are they expected to compete with existing commercial operations. As a result they find it extremely difficult to build up a cash flow which might be used to sustain jobs and purchase equipment when CP funding stops.

Schemes involving small groups of CP workers which are attempting to launch themselves as a new enterprise would be enormously helped if the workers involved were able to move, on completion of the CP, directly onto the EAS which would financially underpin them for a further twelve months and allow them to generate and retain profits. At present this is not possible. To qualify for the EAS individuals must be registered unemployed and in receipt of benefit. A more specific problem restricts the use of the EAS by community business initiatives. Although these enterprises have a long-term commitment towards democratic management and collective ownership, for the purposes of the EAS community business workers are seen as 'employees' not 'entrepreneurs' and therefore do not qualify for the scheme.

Some projects have, often quite ingeniously, overcome these problems. One community business in Glasgow which operates as a large CP managing agent has succeeded in spawning three separate commercial operations from the CP base. One of these operations is a graffiti cleaning service. This began as a CP scheme with equipment provided by the SDA and provided a free service to its sponsors and the local community. During the year of CP funding the cleaning process was refined and developed. In addition, a contract was secured from a national clearing bank, the work for which was carried out at weekends on a commercial basis. This did not appear to breach MSC rules, and the income was retained by the community business. At the end of the twelve-month period of CP funding the scheme had an efficient and tested product, a skilled work-force, and capital back-up, and was almost ready to be launched as a commercial operation.

Clearly, attitudinal and cultural barriers and the debilitating effects of long-duration unemployment on people's skills and aptitudes pose considerable problems in expanding the participation of disadvantaged workers in self-managed economic activity. Prior to any formal self-employment training, some form of developmental and remedial work will be needed to raise awareness about self-employment and to enhance confidence and aptitude. This work may well have to be highly staff intensive and take place over a lengthy period of time.

While the main thrust of MSC policy is not in this direction, initiatives operating relatively independent from MSC resources have been able to combine a variety of pre-training developmental work, longer, flexible, and often informal training inputs, and post-start-up support. One project in Liverpool concentrates on unemployed young people, many of whom begin with only a vague interest in self-employment. After informal introductory sessions with project staff at which the young people decide for themselves whether they wish to continue with the idea, those that do are allocated to an individual member of the project's staff. The two meet regularly on an open-ended basis to develop the business idea. After six months, most participants are able to benefit from a short formal training course which refines the business plan. Following this, specialist courses can be set up to meet individual client needs, partly through tapping the experience of local private sector contacts—accountants, lawyers, marketing specialists, and so on. The project has received funds under the ESF. It uses these, not to pay trainee allowances as such, but to reward individuals who undertake training directly related to improving the chances of successful business ventures. Only when individuals have successfully developed a business plan is a training award made, this payment being conditional on any further training thought necessary being undertaken. Over the first 18 months of its operation the project allocated around £37 000 in grants to forty-four people who had set up twenty-nine companies providing work for seventy-five people. In total the project had 230 people registered with it who were in the process of developing business ideas or already trading.

Another innovative scheme which also attempts to take people with little idea about self-employment to the stage where they are trading and have access to ongoing training and support began in Dundee in October 1984. This initiative—the Dundee Training for Employment and Enterprise Project ('Work Start')—focuses specifically on one of the town's deprived council house estates and operates from premises located in that community. Again, the ESF provides resources: £445 500 has been committed over a three-year period as part of the ESF innovative project programme. Matching funds are provided by the local authority, the SDA, and the Community Projects Foundation. By December 1986 the project aims to provide 330 local people with business training and guidance and to help almost 250 people move into self-employment.[13]

Despite the limitations attached to mainstream MSC self-employment programmes, a number of developments which are potentially supportive of entrepreneurial innovation have taken place within MSC policy. The funding of pilot training schemes for EAS participants and the availability of a basic self-employment training package targeted at the bottom end of the market which can be used flexibly by local trainers are potentially beneficial developments for levering unemployed people into self-employment which have taken place within the MSC.

A further pilot has been introduced to the Voluntary Projects Programme (VPP). Since 1982 the VPP has been supporting projects designed to help unemployed people improve their job prospects through participation in voluntary activities. In 1985 this programme developed an experimental self-employment component. It means that projects which help prepare unemployed people for self-employment and provide initial advice and support to their new businesses can now be funded under the VPP. As a result of these changes, it has been possible for organizations to fund staff under the programme to perform both outreach and after-care functions in the generation of new enterprises. It should, therefore, be possible to stimulate interest within local communities about the possibilities of self and co-operative employment, establish close contact between project staff and unemployed people in an animation and training process, and enable contact to be maintained following business start-ups. One early example of a project tapping the VPP in order to help unemployed people prepare for self-employment is Project Fullemploy's initiative at Sandwell, Birmingham. A resource centre has been set up with £37 000 from the VPP to provide counselling, practical assistance, and information to unemployed people and to those engaged in setting up or running their own businesses. Although the resources are modest—£75 000 being the maximum funding available for any one project in 1985—utilizing the new scope under the VPP may increase the chances of more disadvantaged groups making a success of a new business start.

A capacity for enterprise development also exists within two aspects of the YTS: namely ITeCs and Training Workshops. The production of goods and services to customers on a commercial basis is inherent to the work of these facilities, and both are required to generate a proportion of their own income to survive. In both these structures YTS participants may be involved in activities which have the potential to be developed commercially and launched as new enterprises.

In 1985 a two-year project was launched with funds from the MSC as a pilot exercise to test this potential for creating new businesses. Run under the auspices of the Training Workshop Resource Unit, the Genesis Project builds upon previous efforts to assist Training Workshops to market goods and services produced by trainees. Over the first year of Genesis some twenty projects with the potential to exploit new markets will be selected from ITeCs and Training Workshops throughout Britain. Each of these projects will involve between five and ten YTS trainees. The trainees will continue to receive training in the standard YTS core skills, and training in business skills will be added. Schemes covered by the programme will include: the production and marketing of good-quality Welsh toiletries and slate products; the production of high-quality Scottish knitwear aimed at the US market; and the development of a computer-based tourist information service for hotels. The aim is to develop and launch schemes like these as new enterprises or to attach the product and its trainees to an existing commercial operation. A small number of other projects have also been attempting to use the skills gained during YTS training as the base for a new enterprise. In Somerset, for example, the Foundation for Arts and Crafts Enterprise

(FACE) provides YTS-based training in a range of local craft skills. Again, training in business skills is added, and trainees are encouraged to establish individual or co-operative enterprises on completion of the YTS with FACE providing ongoing advice and support.

With most of government financial provisions for self-employment training tied up with the short formal training courses of the MSC, projects wishing to reach disadvantaged client groups will often have to rely on people partici- pating under their own steam without any special training allowance paid to them. Some of the schemes described above are examples of this. The clients of these projects will be mainly unemployed people, many of whom will be in receipt of unemployment or supplementary benefit. This brings the work of these projects directly into touch with the operation of the social security sys- tem and may well create problems by way of loss of benefit for people involved in innovative employment and training initiatives. Organizers of local projects have often found it useful to keep local Department of Health and Social Security (DHSS) managers informed about the project's oper- ations, sometimes by inviting them to visit the project and talk to staff and participants. Securing the goodwill of the local DHSS and reassuring them that rules are not being broken (for example, that participants are genuinely available for work) is felt to have eased the problems benefit-recipients might otherwise have had.

It may also be possible to go further and secure specific concessions from the DHSS of a financial nature. A fundamental constraint on benefit-recipi- ents engaging in economic activity concerns the amount of money they can receive and retain. The benefit disregard rules are very severe. People in receipt of supplementary benefit can earn up to £4 per week. Anything above is deducted pound for pound from their benefit payment. For people with no savings and virtually no chance of securing normal employment, this makes it extremely difficult to raise the necessary capital to finance a small business start. An initiative in Medway, Kent, has partly overcome this problem by securing a special arrangement with the local DHSS. Unemployed young people can engage in commercial services, such as site clearing, gardening, decorating, and repair jobs that are too small for con- tractors and too big for property-owners. Alternatively, they can begin to work towards creating their own business. The income generated by these activities is held in trust as money credits for the workers by the project. Benefit payments are not threatened so long as individuals remain available for work and do not work for more than 29 hours per week. The cash raised can eventually be used to finance the establishment of a small business or co-operative, or, if people move into employment, the money credits can be cashed and counted as gross income for tax purposes. The DHSS reported in May 1985 that other similar schemes were being proposed in Barnstaple, Torquay, Peterborough, Sheffield, Exeter, Walsall, Dumfries, Lewisham, Cardiff, and Devon.

Lessons and recommendations

In this review of innovations in local manpower policy we have presented a picture of great diversity in terms of activities, the agencies involved, and the client groups affected. There is clearly great strength in this diversity and in the growing body of expertise among those involved at the local level. To conclude this section we shall now highlight some of the key factors which appear to be constraining what might be achieved by creative local action. These observations relate to statutory and financial issues and to certain technicalities attached to the operation of existing programmes. We also set out a number of practical suggestions for improving the possibilities surrounding local innovation. Short of fundamental resourcing changes, we feel the scope for innovation would be enhanced if greater flexibility and integration could be introduced to existing policy provisions and if more progress was made in the area of local labour-market planning.

In many of the initiatives we have discussed, local authorities play an important role as initiators and resourcers. However, in the light of the recent experience of the GLC, it may in future be difficult for more authorities to develop independent local labour-market strategies because of the lack of statutory responsibilities in this area and the tight expenditure controls exerted by governments. If government were to afford local authorities more legislative powers to intervene in local labour markets, as it has done in the area of local economic development, the potential for local authority action could be markedly improved. Legislative innovation of this sort would also allow greater co-ordination between manpower and economic development programmes at the local level. This is an area where little headway has been made so far.

We have noted that many innovative projects, while locally devised, are often dependent to some extent on centrally allocated resources such as MSC programmes, Urban Aid, the ESF, and even social security payments. Feedback from projects which tap these resources indicates that the rules and regulations which govern these programmes can have an important bearing on the potential client group and the scope for innovation. Eligibility criteria determine who can participate on a scheme, and in certain cases groups of disadvantaged workers can be excluded. Key MSC programmes, like the CP and the EAS, are targeted exclusively at the unemployed in receipt of benefit. While this guarantees that resources will flow to a large client group in particular need, the real level of unemployment, and the hardship it causes, is greater than the official unemployment total. Many people are actively looking for work but do not register as unemployed because they are not entitled to benefit. A large proportion of these are women who have not paid full national insurance contributions in the past. As was pointed out earlier in the chapter, this client group has recently been excluded from CP eligibility as a consequence of changes to programme regulations.

Other constraints on innovative practice which stem from the rules attached to main programmes relate to how these programmes can be used by local projects. With particular reference to community businesses, we have shown how schemes which utilize training and job creation programmes frequently find it difficult to get involved in commercial activities, and as a consequence can be restricted in their development potential. Some local initiatives have also experienced difficulties in stitching individual programmes together to form an integrated package of support. In the last section on generating new economic activity we identified the current difficulty of bringing together the CP and EAS within community business structures, despite each element having roughly the same client group in mind. If such integration were possible, it would also increase the scope for complementing the CP and EAS with other programmes such as Urban Aid and the ESF. Some local initiatives are managing to make progress in the area of integration and cross-fertilization of existing resources. The METEL training facility we described earlier from Liverpool is one good example of such an approach.

It would appear particularly timely for more consideration to be given at the policy level to the scope for programme integration. Lasting large-scale deficiency in labour demand has created a substantial client group of very long-term unemployed. In January 1985, for example, almost 800 000 people had been unemployed for two years or more. There are limits to what participation on individual programmes can do for this client group. To progress these workers to a position from where they can compete effectively for jobs, contemplate establishing and managing a new enterprise, or engage in rewarding forms of work of a less conventional kind, a more programmatic approach will probably be required. To cater for this group, manpower strategies will be needed which can move the individual through a sequential process involving a variety of programme and support measures sensitively designed to counter the damaging effects of long-duration unemployment, and which can combine developmental and remedial elements. There is clearly a trade-off involved, as more help for one group means less for others at whatever level of resourcing applies. However, we would argue that the problem of long-term unemployment is such that more resource-intensive approaches are required if there is to be a reasonable chance of success.

As we have shown in the discussion of local manpower initiatives, the main innovations of this sort appear to be taking place at the level of individual local projects. There seems to be insufficient commitment to an integrated approach within the overall design of national manpower policy. In part this is an understandable consequence of the rapid expansion of MSC special programmes since the late 1970s, introduced on an *ad hoc* basis in response to the escalation of national unemployment. Nevertheless, when one looks at the range of existing policy provisions we have described in the body of this chapter, it should be possible to develop a more programmatic approach. For example, at one end of the range there is the possibility of working with unemployed people in an open-ended and flexible way under the VPP. At the other there are the formal training programmes under the ATS which seek to link people with jobs or new business starts. Although progression between each different element is possible, the position of the long-term unemployed could be improved if more well-defined paths were laid down. This is something which could perhaps be taken forward on an experimental basis.

So far we have discussed a number of constraints on the development of successful employment and training initiatives at the local level. In many instances the removal of these constraints would imply the commitment of extra resources to this general area of activity. To bring the discussion to a close we now give detailed consideration to the most effective mechanism for deploying existing and future resources to meet manpower needs at the local level. In effect, we argue that there is a strong case for a greater degree of *local labour-market planning*. We begin by setting out the basis for this case, move on to an examination of the substantive content of a local labour-market planning function, and finish by looking for an appropriate location for such a function.

At the beginning of the chapter the dramatic rise in national unemployment was noted. This has induced a major growth and development in national manpower programmes, and evidence was presented on this from MSC budgetary information. This represents a nationally planned response to a serious national problem. For the implementation of a number of the new programmes the MSC has depended on a large number of bodies outside of their own organization. These include voluntary bodies, local authorities, and private sector employers, as we noted in earlier sections. Additionally, in a number of urban areas, manpower initiatives have been developed which do not depend on MSC funding. The EEC, the local authorities, and the voluntary sector have played important initiating, funding, and managing roles here. The upshot is that in many towns and cities there is now a scale and diversity of manpower initiatives which was undreamt of in the late 1970s.

The basic case for some degree of local labour-market planning can be located in these developments. First, the MSC in developing the scope and content of its programmes is responding to national imperatives. However, these national programmes need to be implemented in local areas with very different problems and needs.[14] For example, at an earlier stage of this chapter the differential urban impact of training measures oriented more towards the needs of employers than the unemployed was highlighted.

Second, partly in response to the inadequacies and inappropriateness of some elements of national provision, a multitude of local initiatives have developed. In many instances these have been tailored to priorities set in Brussels and not the local area. But in any event they flow from a wide range of organizations, and there is a clear need for a more planned approach. An additional consideration is that most EEC and MSC programme monies are released on an annual basis. Scheme sponsors and agents are operating in a situation characterized by uncertainty and short time horizons. Although more secure funding arrangements are the solution to this difficulty, local co-ordination and planning of activities may help by situating individual initiatives within the context of a longer-run attack on local labour-market problems.

We now consider the possible functions of a unit concerned with local labour-market planning. This discussion is carried out in such a way as to bring forward a number of more detailed justifications for this type of approach to local manpower problems and the programmes designed to cope with them.

The first element involved in local labour-market planning would be the preparation and updating of a statement on the manpower and training needs of an area. This would include the establishment of priorities within the local area. There is a sense in which this is done nationally using fairly crude criteria, such as the numbers of long-term unemployed in a particular region. Clearly, however, this can be most effectively carried out by a local body. Various interests in the local economy and community could present their case for priority treatment.

Secondly, given a clear, locally generated set of priorities, the local labour-market planners would be in a strong position to respond to proposed changes in national manpower policy. Changes in MSC programmes of a major nature are generally preceded by a process of consultation. A well-resourced local body with a well-developed understanding of the local labour market and a clear set of local priorities would be able to assess the likely impact on their area of proposed changes in national policy.

Thirdly, there is an ongoing need to monitor local manpower provision and to match this up against local needs. The next step is to identify gaps in provision, or programme areas which require expansion or contraction in response to changes in the nature of local problems. As labour-market problems have grown and provision has expanded and diversified this particular task has become increasingly difficult to carry out. The complex and fragmented nature of provision in urban areas with a high level of local manpower initiatives can be effectively monitored only by a locally based body. Regular monitoring of provision in relation to needs could permit a local labour-market planning unit to take an active role in the development of new initiatives. It could identify specific areas of need (for example, particular locations, labour-force groups, or programme areas) and seek to organize a practical response from the MSC, the local authorities, the voluntary sector, or other relevant potential providers of schemes.

Fourthly, an important function of the local unit would be to enhance the possibilities for co-ordination and cross-fertilization of manpower initiatives and services in its area. We have already described the functions of stating priorities, evaluating provision in relation to need, and identifying gaps in provision. The issue of co-ordination has become important because of the increased spread of manpower programmes and schemes available and the growing number of organizations involved in their delivery. The MSC's Local Collaborative Projects represent an interesting experiment in bringing together on a local basis the providers of training and one component of the demand for training—the employers. However, there is a need for a more comprehensive approach to co-ordination at the local level. The major benefits of this may lie in the generation of stronger links with other major activities such as economic development and housing. A single local body with expertise straddling all aspects of the manpower field would be in an ideal position to link its programmes with a more general local economic development initiative.

If there is an acceptance of the general case for a greater degree of local labour-market planning, there remains the issue of a suitable location for the body or unit which would carry out the planning. In the absence of a satisfactory research base it is difficult to answer this question. It is possible to conceive of various locations, including local authorities, voluntary sector organization, the MSC , and so on. Some local authorities could clearly carry out such a function, and have developed interests and expertise in the manpower area. However, a recent study of Merseyside's manpower services (Dabson *et al.* 1984) argued the case for a local labour-market planning unit based on the local Area Manpower Board (AMB).

The AMBs are part of the basic organizational structure of the MSC and represent a wide variety of local interests—education, industry, trade unions, and voluntary groups. The Merseyside study suggested that some of the functions involved in local labour-market planning were part of the remit for AMBs. However, the rapid development of MSC main programmes such as the YTS and CP had placed heavy administrative burdens on the Merseyside AMB. Additionally, the board lacked a secretariat which could provide detailed research and statistical and other relevant briefing material. It was felt that the creation of such a back-up resource would allow the Merseyside AMB to plan the local labour market.

There were two key advantages to the AMB as opposed to other possible administrative locations. First, the AMB was representative of a wide spread of local opinion and had members drawn from the main relevant organizations. Second, the AMB had direct access to the principal agent of manpower policy in the area—the MSC. This having been said, the Merseyside situation is not necessarily replicated in other urban areas. One key factor is the *active* involvment in the AMB of a number of locally well-known and well-respected individuals. This lends authority and credibility to the board—qualities which would be required for any body exercising an important planning function. It may, therefore, be dangerous to generalize too far from the Merseyside situation. It could be argued, of course, that, by resourcing the AMBs to permit them to carry out a local labour-market planning role with greater effectiveness, existing members would become more actively involved and good-quality new members become easier to recruit.

Concluding Remarks

In the last section of the chapter we tried to summarize some of the main points arising from the study and make a number of positive suggestions for improving the effectiveness of local employment and training initiatives. As much of the discussion centred on what was happening at the local level, it is all too easy to lose sight of the broader economic environment. We conclude with two general points which serve to underline the importance of such a depressed environment.

The first point concerns the effectiveness of the MSC. Operating as it does in an extremely depressed economic environment, it is not surprising that many MSC programmes attract criticism from national and local organizations. However, fundamentally, many of the problems associated with MSC programmes reflect the economic situation and not the inadequacies of the organization itself. This applies particularly to the low placement rates from programmes like the CP and the only modest placement rates from others like the YTS. Additionally, as the government's front line organization in the unemployment arena, the MSC has experienced dramatic growth in programme budgets and has been subjected to regular and substantial changes in programme mix and content. The YTS is a classic case, having scarcely come of age before major changes in the length and format of the scheme were brought forward. The irony here is that in the same sense in which organizers of local initiatives and sponsors of CP and YTS schemes complain about short time horizons and annual funding, MSC operational staff might equally complain about the uncertain and ever-changing environment within which their own organization is forced to live.

The second point is that the scope for local manpower initiatives is constrained by the generally depressed state of the economy in the same way that national programmes are constrained. While these local initiatives remain small in scale, it will be possible for them to show good results. Were they to be expanded significantly it is likely that diminishing returns would set in. This applies particularly to schemes which are attempting to train people for existing jobs. The irony is that, whereas economic depression has stimulated the growth of local initiatives, a degree of economic recovery is required before they can begin to make a major impact.

Notes

1. Detailed analysis of the Bristol, Glasgow, London, and West Midlands labour market can be found in the various case-study volumes of the Inner Cities

Research Programme. Other urban areas can be analysed by looking at local authority or consultancy reports. For example, Merseyside's labour-market problems are evaluated in Roger Tym and Partners (1984).

2. On the other side of the coin, however, the role of the Industry Training Boards has been much reduced and the higher education sector severely cut back.

3. The White Papers were *Education and Training for Young People* (HMSO 1985b), and *Employment: The Challenge for the Nation* (HMSO 1985a).

4. There is, however, still a fundamental question as to the actual significance of training in economic development. Space does not permit an analysis of this here. It is perhaps not unfair to suggest that the government's message is based on casual empiricism rather than a thoroughgoing analysis.

5. Around 80 per cent of places are now part time, although the original Job Creation Programme was full time. Nevertheless, the budget commitment has expanded greatly both in real terms and as a proportion of MSC spending.

6. Here the net cost is calculated by subtracting Exchequer flowbacks (lower benefit payments, higher tax and national insurance receipts) from the gross cost.

7. This growth has to be seen in the context of the demise of MSC funding support for the Industry Training Boards. This support formerly included substantial inputs into trying to sustain a reasonable level of apprentice training. In this light, the growth of effort in the youth training field is less marked. However, first-year apprenticeships are fairly widely supported within the YTS.

8. The TVEI is an experiment which is seeking to develop technical and vocational education for the 14–18 age group. NAFE refers to the plans to give the MSC more leverage on non-advanced further education so that it might be geared more directly to the needs of local employers.

9. In Knowsley in Merseyside, 90 per cent of Mode A places were reported to be provided by private training organizations in 1984 (Roger Tym and Partners 1984).

10. The funding of occupational training for the unemployed must actually fall to pay for the introduction of local grants to employers and other adjustments to the adult programme.

11. Two of the twenty-nine Skillcentres were reprieved, both in the South of England (Twickenham and Southampton). Of the five Scottish centres closed, four were in the economically depressed Glasgow conurbation (Hillington Annex, Queenslie, Dumbarton, and Bellshill Annex).

12. In 1985–6, for example, the TEP budget of £14.5 million was expected to cater for around 12 000 new starts and 4 000 existing businesses.

13. Dundee Training for Enterprise and Employment Project, *Inaugural Report*, Oct. 1984.

14. A strong case for greater co-ordination of MSC services at the local level was made in a report on Merseyside carried out in the 1970s (MSC 1977). The report concluded that 'There is a clear need for the differing activities of manpower services to be co-ordinated and it is essential that focal point should be on Merseyside . . . so that it can be responsive to local needs and priorities, and to changes in local circumstances' (para. 8.2).

References

Brenner, M. H. (1979), 'Mortality and the National Economy: A Review of Experience in England and Wales, 1936–76', *Lancet*, ii. 568–73.

Buck, N., Gordon, I., and Young, K., with Ermisch, J. and Mills, L. (1986), *The London Employment Problem*, Oxford: Clarendon Press.

Dabson, B., McGregor, A., and Collins, M. (1984), *Manpower Topic Review*, Liverpool: Centre for Employment Initiatives.

Department of Employment (1984), 'Evaluation of the Pilot Enterprise Allowance Scheme', *Employment Gazette*, Aug. 374–7.

Department of Trade and Industry (1984), *The Human Factor—The Supply Side Problem*, IT Skills Shortage Committee, First Report.

Department of Trade and Industry (1985), *Changing Technology—Changing Skills*, IT Skills Shortage Committee, Second Report.

EEC (1983), *Community Action to Combat Unemployment. The Contribution of Local Employment Initiatives*, Brussels: EEC.

Forbes, J. F. and McGregor, A. (1984), 'Unemployment and Mortality in Post-War Scotland', *Journal of Health Economics*, 3, 239–57.

GLC (1985a), *Greater London Training Board, Annual Report, 1984*.

GLC (1985b), *Charlton Training Consortium—Review of Progress and Funding for 1985–6*, Report 20.3.85 by Director of Industry and Employment and Chief Economic Advisor.

Goodman, J. F. B. and Samuel, P. J. (1966), 'The Motor Industry in a Development District', *British Journal of Industrial Relations*, 4, 336–65.

Gravelle, H., Hutchinson, G., and Stern, J. (1981), 'Mortality and Unemployment: A Critique of Brenner's Time Series Analysis', *Lancet*, 26 Sept., 675–9.

HMSO (1984), *Training for Jobs*, Cmnd. 9135, London: HMSO.

HMSO (1985a), *Employment. The Challenge for the Nation*, Cmnd. 9474, London; HMSO.

HMSO (1985b), *Education and Training for Young People*, Cmnd. 9482, London: HMSO.

McArthur, A. A. (1984), 'Public Responses to the Growth of Unemployment in the United Kingdom with Particular Reference to Action at the Local Scale', unpublished Ph.D. thesis, University of Glasgow.

McArthur, A. A. and McGregor, A. (1986), 'Policies for the Disadvantaged in the Labour Market', in W. F. Lever *et al.*, *City in Transition*, Oxford: Oxford University Press, Ch. 8.

Mason, C. (1983), 'Labour Market Policy', in K. Young and C. Mason (eds.), *Urban Economic Development, New Roles and Relationships*, London: Macmillan.

Middleton, A. (1981), 'Local Authorities and Economic Development', *Centre for Urban and Regional Research, Discussion Paper No. 1*, University of Glasgow.

MSC (1977), *Manpower Services on Merseyside*, Report of the Merseyside Steering Group.

MSC (1980), *Outlook on Training. Review of the Employment and Training Act, 1973*.

MSC (1981), *A New Training Initiative. A Consultative Document*.

MSC (1983), *Towards An Adult Training Strategy. A Discussion Paper*.

MSC (1984a), *Survey of Community Programme Participants*, Employment Division, Aug.

MSC (1984b), *Community Programme Postal Follow-Up Survey*, Employment Division, Dec.

MSC (1985), *Labour Market Quarterly Report*, May.

NCVO (1984), *Joint Action—The Way Forward*, London: Bedford Square Press.

NEDC (1984), *Competence and Competition*, London: National Economic Development Office.

Pilgrim Trust (1938), *Men Without Work*, Cambridge: Cambridge University Press.

Project Fullemploy (1984), *Annual Report 1983–1984*, London.

Roger Tym and Partners (1984), *Merseyside Integrated Development Operation. Manpower Training Review*, London.

SCB Ltd. (1985), *Vocational Training Programmes 1984, Interim Report and Evaluation*, Feb., Glasgow.

Warr, P. B. (1983), 'Job Loss, Unemployment and Pyschological Well-being', in V. L. Allen and E. Van De Vilert (eds.), *Role Transitions*, New York: Plenum.

6

Housing Policies, Markets, and Urban Economic Change[1]

John Ermisch and Duncan Maclennan

Introduction

The focus of the Economic and Social Research Council (ESRC) Inner Cities Research Programme has been on the poor job prospects faced by many inner city residents and the further effects that a concentration of poor people may have on other inner city residents who have jobs and are not living in poverty. This essay considers how housing policy may influence the extent of these inner city problems. In particular, it considers the effect that housing policy can have on job creation in the inner city, on the concentration of poor people there, and on the ease with which inner city residents can move to areas having better employment opportunities. Special attention is given to the ways in which housing policies affect the social mix of neighbourhoods, their attractiveness for investment by households and firms, and the mobility of households.

For almost a century a close connection between the quality of urban life and housing conditions has been recognized in Britain. However, the nature of the connection emphasized in housing policy has varied from time to time. In the late nineteenth century, basic forms of housing policy (building controls and demolitions) were seen to be essential in order to expand the supply of urban labour and to make cities liveable at even a barely tolerable quality of life. Half a century later, and indeed continuing through to the early 1970s, large-scale housing programmes with narrowly defined shelter objectives characterized British conurbations and were usually pursued independently of any assessment of their local economic impacts. We shall argue below that the central features of British housing policy, which in retrospect were similar in design for most governments until 1979, have not always made a positive contribution to urban policy.

At first sight, the White Paper which inaugurated an era of official concern for 'inner areas', *Policy for the Inner Cities* (HMSO 1977), appeared to promise an integrated description of the problems and analyses of the causes of inner city decline. Attention was drawn to the prevalence of poverty, poor environment, and bad housing conditions. All of these symptoms of the problem revolve around the spatial concentration of economically disadvantaged people, particularly the unemployed. The potentially cumulative nature of this disadvantage was appreciated. That is, the concentration of the unemployed and those facing a high risk of becoming unemployed in inner cities reduces their chances in the labour market. They live near the centre of labour markets which have lagged behind less urbanized areas in creating employment opportunities for a long time, and their concentration there means that each person faces a large amount of competition for available jobs. But the White Paper argues that the concentration of poor people has further effects which go beyond the people experiencing unemployment and poverty:

... there is a collective deprivation in some inner areas which affects all the residents, even though individually the majority of people may have satisfactory homes and worthwhile jobs. It arises from a pervasive sense of decay and neglect which afflicts the whole area, through the decline of community spirit, through an often low standard of neighbourhood facilities, and through greater exposure to crime and vandalism which is a real form of deprivation, above all to old people. All this may make it harder for people to maintain their personal standards and to encourage high standards in their children. Sometimes people from particular inner areas experience extra difficulty in getting a job or a mortgage. This collective deprivation amounts to more than the sum of all the individual disadvantages with which people have to contend. (HMSO 1977, 4.)

Taking the above points in turn, it is clear that the 1977 White Paper recognized the complexity of the problem, that specific area problems were largely determined by city-wide economic processes of decline and concentration mechanisms, and that 'externalities' or 'environmental side effects' were an important consideration. Regarding the latter point, housing conditions, neighbourhood quality, and area image (or confidence) were all, to some extent, joint products of housing investment.

However, this conceptual framework, if adequate for encapsulating how areas have declined, needs to be expanded and revised if we are to understand the series of channels by which housing spending and policies may have an impact upon urban economies. First, we believe that it is helpful to drop the rather casual geography implied in the term 'inner areas' or 'inner city', as this implies little about the scale or characteristics of areas which policy might address. Instead, it is helpful to conceive of the city as constituting a series of neighbourhoods of varying quality. In consequence, deprived or declining neighbourhoods may be in the public or private sector, and they may be on the edge of the city as well as in or near the centre. Second, while accepting the emphasis on the declining economic base as the key problem at the metropolitan level and its implied causality running from economic change to housing investment, we stress that regeneration policy also has to seek out the linkages which run from housing and environmental investment to local area economic vitality. The inner cities White Paper suggests that:

This shabby environment, the lack of amenities, the high density remaining in some parts and the poor condition of the older housing in the inner area contrast sharply with better conditions elsewhere. They combine together to make these areas unattractive, both to many of the people who live there and to new investment in business, industry, and housing. (HMSO 1977, 3.)

Thus, if the collective deprivation and externalities referred to above really do exist, then some critical minimum effort of housing reinvestment may be required to promote net economic growth for a large urban tract or even the metropolitan area as a whole. These longer-term economic effects have to be distinguished from the short-term employment and multiplier effects which flow from housing reinvestment.

Third, housing policies affect the degree and direction of household mobility. By doing so, they affect the distribution of socio-economic groups (SEGs) across a metropolitan area. Housing policies thereby influence the concentration of poor people in particular areas, which in turn may affect the economic prospects of these people. We do not believe that these critical issues for policy have yet been researched in relation to the 'Inner Areas' programme, and they still receive negligible attention in the most pertinently related heading of housing policy—that is, the area-related housing renovation grant programmes. In this chapter we try to draw together the rather scanty evidence which exists on these points.

The main aim of this chapter is to set out the economic influences which concentrate problems into specific areas of cities, the problems of efficiency and economic mobility which may arise, and the ways in which housing policy, or at least its components, either enhances or frustrates broader policies for urban economic development. To this end, the rest of this essay is structured into four major sections. First, we examine the mechanism by which concentration of disadvantaged households arise and their broad spatial patterns. Second, we briefly examine the housing–economy interactions which can arise with housing reinvestment in existing neighbourhoods. Third, we present a cameo description of recent trends in British housing policy, which still seems to be developed independently of urban policy. Fourth, we consider in some detail the impact of specific areas of policy upon particular kinds of neighbourhood and, in turn, the urban economy. Since some major areas of housing policy are now under review, we also include suggestions for policy reform. Finally, there is a short conclusion.

Separation of SEGs

A plethora of studies, largely based upon census data, of the residential spatial structure of North American and British cities were undertaken in the 1960s. Even though there are few more recent British studies, and none which examine income separation by small area (as income data are not recorded in the Census), the broad findings of such 'factorial ecology' studies probably hold as a reasonable description of the spatial structure of British cities. These studies emphasized, within the private housing market, the separation of different SEGs into different neighbourhoods. A broad ring structure of progressively higher-status SEGs was observed, even if the overall structure was much disrupted by enclaves and sectors. And, of course, in British cities the non-market sector, often constituting as much as half of the housing stock by the 1970s, generated its own pattern of neighbourhoods and, somewhat unexpectedly and rapidly, socio-economic separations. In this section we discuss market and non-market sectors separately.

The market sector

Urban economists, though more concerned with population density gradients and rent/land value gradients, were quick to indicate that the observed spatial structure of the market sector was quite consistent with the access–space trade-off model of residential location choice. That is, households can purchase more space (larger housing) more cheaply by moving away from central city locations, but only if they are willing to incur higher direct and indirect (cost of time) commuting costs. The conventional wisdom, which has rarely been tested in the UK, is that a higher income increases the demand for space by a greater proportion than it increases the demand for access. Hence, higher-income households live on larger properties further from the city centre and the poor live at high density in more valuable central city land.

While this model may be a useful meta-theory for examining some long-run aspects of spatial structure, it is relatively unconvincing as a model of spatial choice at a particular point of time. This limitation arises for a number of reasons. First, the aggregative nature of the model directs our attention away from the variety which exists, either in house or household types, at any given distance from the city centre. Neighbourhood regeneration is much concerned with variety rather than uniformity—standard errors are as important as means. The patterns in the map in Figure 6.1, which groups 1981 Census Areas for Glasgow into 'housing product groups', indicates that there is no simple distance–dwelling type of relationship, and at the same time the map casts doubt upon any notion of a simple 'inner city' set of problems. Thus the model is only consistent with the facts if we make the facts very stylized, whereas neighbourhood analysis requires attention to detail.

A further problem arises in that housing economic analysis stresses the variety of characteristics associated with a house. That is, the dwelling choice is not merely a selection of location relative to employment and numbers of space units, but it is also a choice of dwelling style and amenity, access to neighbourhood facilities (public and private) and amenities, and household access to a broad set of activity points in the metropolitan area. This rather

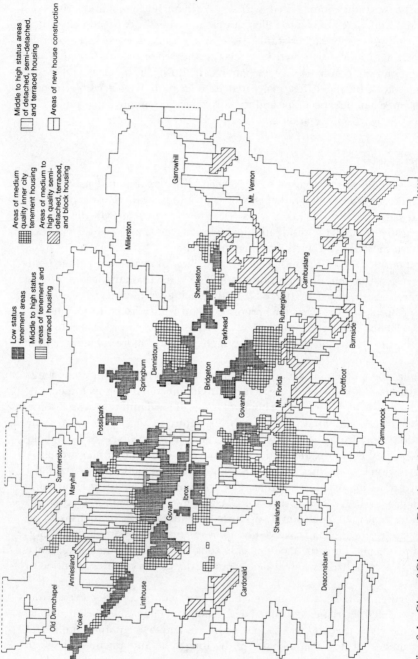

Fig. 6.1. City of Glasgow District: Private Sector Housing Areas.

Note: The housing and neighbourhood groups are based on cluster analysis of 1981 Census data.

Sources: Glasgow University Database and 1981 Census.

Table 6.1. *Average Prices of Houses in Radial Bands round Central London: Three-bedroom Houses sold between October 1965 and March 1966* (Amount in £)

Radial bands: Approx. distances from Central London	Second-hand houses				New houses	
	Built before 1919		Built 1945–64			
	Terraced	Semi-detached	Semi-detached	Detached	Semi-detached	Detached
41–60 miles (beyond the Outer Metropolitan Area)	2 840	3 470	4 340	5 380	3 980	5 670
21–40 miles (most of the Outer Metropolitan Area)	3 100	3 730	4 730	6 130	4 530	5 930
13–20 miles (the fringe of Greater London)	4 170	4 680	5 640	7 370	5 570	7 700
3–12 miles (Greater London outside the centre)	4 550	4 930	5 770[b]	9 790[a]	5 690[a,b]	7 630[b]
The centre (Westminster, Camden, and Kensington and Chelsea)	11 760	14 630[a]	—[c]	—[c]	—[c]	—[c]

[a] Derived from a small sample.
[b] Majority of sample is near the 12-mile limit.
[c] Inadequate sample for figures to be given.
Source: Greater London Council (1968, 153).

obvious, and well-confirmed, set of observations regarding housing as a commodity has quite important implications for the prospects for neighbourhood regeneration.

The pure form of the access–space model always implies a rather grim future for core neighbourhoods—namely, that income growth for a household in such locations inevitably implies decentralization with only the poor remaining. Again an initial examination of available data suggests that house-price gradients (see Table 6.1 for London and Figure 6.2 for Glasgow) do appear at the overall metropolitan level. There is also evidence to suggest that better-off households have tended to move into new suburban housing, while slightly less well-off households have moved into the housing vacated by them. Most moves of house have been over short distances and have been motivated by housing reasons (see Gordon *et al.* (1983) for the London region), and moves in the owner-occupied market appear generally to entail an improvement in the quality and/or size of housing (Jones 1978). Thus, the adjustment to better housing and neighbourhoods appears to be gradual, with dwellings passing to lower income groups as each income group consumes better housing. Jones' (1978) study of movement by owner-occupiers in the Manchester area during the early 1970s found that, despite the general tendency to increase housing consumption upon movement, three-fifths of

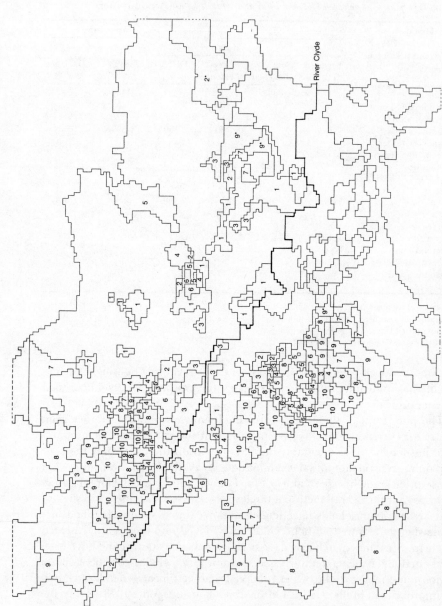

Fig. 6.2. City of Glasgow District: House-price Deciles of Census Areas, 1983.

Note: The asterisked figures are for areas with less than ten transactions in the year.

the sample paid less for their present house than they received for their previous house, which suggests that they moved outwards where the price of equivalent housing tends to be lower. A study of movement in the Manchester area in the late 1970s by Hedges and Prescott-Clarke (1983) found that moves within the inner city and out of it generally entailed a movement into better housing, and that housing reasons were the main motive for the move. Movers into the inner city were generally not motivated by the desire to consume more housing, and they did not generally do so. This study, as well as that by Gordon *et al.* (1983), again points to the importance of increasing housing consumption in motivating movement in a metropolitan area, particularly outward movement by owner-occupiers. Less well-off households entering a metropolitan area or forming their first household have tended to move into the older, more centrally located private housing vacated by those moving to better housing.

In private housing markets, housing quality tends to rise with distance from the centre of the metropolitan area, while the price of housing of a given size and quality tends to fall. The upgrading of housing in response to income growth in a private housing market therefore encourages households to move away from inner areas to suburban areas. This tendency was most apparent in London between the wars, when there was a large private house-building boom concentrated in the suburban areas of the London region. During 1921–37, the population of outer London increased by $1\frac{1}{2}$ million while the population of inner London fell by 400 000. The suburban house-building boom was fuelled by economic prosperity in the region, which not only allowed households to consume more housing, but also stimulated new household formation and migration to the region.

A great deal of this evidence seems to be consistent with a simple access–space model. However, we have to recall that distance from the city centre, age of housing stock, and neighbourhood characteristics are highly correlated with one another. Thus, particularly with jobs also decentralizing to the periphery, it is pertinent to establish whether decentralization of higher-income households has been related to demands for space. Econometric evidence from the USA, and to a more limited extent in the UK (Kain and Quigley 1975; King 1976; Diamond 1980; Pollakowski 1982; Segal 1980; Wilkinson 1973; Maclennan 1985) suggests that it is the qualitative aspects of housing demand which are particularly responsive to higher income ('income elastic'). That is, it is dwelling quality and neighbourhood quality which particularly shape choices as incomes grow. At present we do not have time-series estimates of the income elasticity of demand for different attributes. Cross-section estimates of the overall demand for housing in the UK suggest an income elasticity of between 0.4 and 0.7 (see Maclennan, 1982). Estimates in the USA are somewhat higher, commonly in the range of 0.6–0.9. More importantly, US studies for specific attributes indicate that it is often environmental and neighbourhood factors, such as access to waterside sites etc, which have very high income elasticities. Recent research in Glasgow (Maclennan 1985) also suggests that neighbourhood quality variables are much more income elastic than either internal housing amenity or accessibility.

These estimates suggest that it is the quality of housing and neighbourhoods in run-down areas, rather than their inherent location, which repels higher income groups.

If this array of partial evidence has wider meaning, then the decline of older neighbourhoods is not inevitable and dependent upon changing commuting accessibility. Rather the crux becomes the extent to which housing structures can be altered and run-down neighbourhoods improved. There are possible impediments to upgrading one's existing house, and neighbourhood financial market imperfections which may be important—and unrecognized in a simple access–space model—in preventing such adjustment.

For instance, housing is durable, which may make it costly to renovate a particular dwelling. But, even if housing can be renovated cheaply, it is difficult to improve the dwelling's immediate surroundings. That is, the neighbourhood environment is outside an individual household's control. It is much easier for a household to improve its housing and environment by moving to an area with higher-quality housing and amenities than to transform the area in which it is living. Thus, income growth is likely to induce households to move to better neighbourhoods as a way of consuming better housing.

Low levels of finance availability from major credit sources can preclude renovation by existing and mobile residents, and—whatever the explanation—studies of Glasgow, Aberdeen, Newcastle, and Birmingham (Maclennan and Jones 1984; Jones and Maclennan 1982; Boddy 1976) have indicated that such patterns were prevalent in the mid-1970s. We discuss below how they may now have altered. Naturally, grant aid for renovation, coexisting with inflation in land and new construction costs, may also encourage renovation in older neighbourhoods.

In a similar fashion the building industry, historically attracted by lower peripheral land costs, may also have contributed to income separation and decentralization. Where population and real income are growing there is much sense in builders providing for the upper half (say) of the house-price distributions, as such a strategy minimizes their risks. However, when, as is now the case in a number of northern British conurbations, real income growth is modest, household growth is focused upon young small households, and decline has provided vacant brown-field sites, then new building patterns may emerge. In Glasgow, for instance, some 3000 moderately priced, smaller housing units have been developed, often on brown-field sites in or near to deteriorated neighbourhoods, since 1980. And developers are extensively involved in converting former warehouses and public buildings into small accommodation units. Such a pattern is less likely to appear in urban areas with a still fast rate of economic growth—for instance, London or Aberdeen—where central land is still pre-empted by either non-residential land users or high-density and high-value housing (as in Docklands).

A range of influences, including demographic change, rehabilitation policy, the emergence of brown-field sites, the changing attitudes of lending institutions, and the rising cost of domestic energy and transportation, are all consistent with a renewed willingness of higher-income households to eschew decentralization urges. New supply structures, rehabilitation possibilities, and financing approaches have only allowed a fuller expression of middle- and higher-income preferences during, at most, the last decade. Where this process involves gentrification (the displacement of low-income residents followed by grant-aided improvement and occupation by higher income groups), then the new 'social mix' may create some new policy problems. However, in Glasgow, which was regarded as Britain's most problematic set of older neighbourhoods in 1972, large areas of the older housing stock have housed higher soci-economic status groups without apparent displacement since 1975. Further, the newly constructed brown-field units house a range of SEGs different from their 'host' neighbourhoods, with motive for location choice often being to secure new or higher-quality housing nearer to the city centre (Lamont *et al.* 1985). Up to the present, development has focused upon provision of units for childless households, and this must raise concern for the future. Econometric analysis (Maclennan 1985) indicates that it is the presence or absence of children in a household rather than permanent income which most influences the decision to live in or near run-down areas. As a prelude to further area regeneration, the Centre for Housing Research is currently attempting to establish the propensities of higher-income households with families to move into new larger higher-quality units in renovated neighbourhoods.

We do not wish to labour these issues further. Decentralization and income separation have been nearly, if not entirely, pervasive features of urban residential development in Britain for half a century at least. There are, however, encouraging signs that this process is not, at the margin, irreversible, and that supply and institutional restrictions as much as transport and building costs have contributed to this process since the 1960s. The market may take a long time to develop new confidence and responses. From a research/theoretical perspective, we need to know much more about the processes of housing choices and to resist interpreting *ex post* outcomes against a rather simplistic underlying model.

The non-market sector

In some metropolitan areas, private housing markets have played a smaller role in the decentralization of residences over the past 80 years. Nevertheless, the movement of people to the suburbs characterized all of the major conur-

bations during the inter-war period, accelerating a tendency which was produced by private housing markets before the turn of the century. In all of the
major conurbations, numbers in the centre remained constant or declined
during the inter-war period, while all of the growth took place in the suburbs.
The London region differed mainly in its large scale of suburban migration
and in the domination of the process by private housing markets. In contrast,
the public sector played a large role in suburban development during the
inter-war period in the Glasgow metropolitan area (Maclennan 1981). Glasgow Corporation was responsible for most of the house-building in the inter-
war years, and the new public estates were mainly on the outer edge of the
inner zone (Richardson and Aldcroft 1968). Other conurbations fit in
between these two extremes of public–private involvement in the process of
suburban development.

The other prong of public sector activity was slum clearance, usually of
privately let dwellings in the centres of the conurbations. There were large
amounts of slum clearance during the 1930s, and also during 1956–75. At the
same time, local authorities were building new dwellings to replace the
demolished ones and to add to the stock of rental housing in order to improve
housing standards and cater to the growth in the number of households.
Thus, this activity was the public sector counterpart of housing upgrading in
the private sector, although its character was 'lumpier' and discontinuous. In
the northern cities, a large part of the new building was in suburban estates;
the pattern for Glasgow is indicated in Figure 6.3. In London this was less
feasible because of resistance to local authority housing by suburban communities (Young and Garside 1982), and local authority house-building was
particularly concentrated in inner London (see Figure 6.4).

The development of this large-scale non-market sector in most British
cities has important implications for the relationship between housing consumption and income and the spatial pattern of SEGs. First, after 1954 the
characteristic programmes of municipal public housing were developed on a
large scale, for economies in land assembly and construction. Or, later in the
1960s, they employed large-scale high-rise buildings. This approach guaranteed from the outset that, regardless of the social composition of council tenants, public and private sector residents would be sharply and visibly
separate. Glasgow's peripheral estates, for instance, aside from their incon-
venient locations all share a monotony of repetitive design, high residential
densities, and an absence of the mixture of land uses and amenities which
characterize older neighbourhoods. Until the 1970s, however, such design
inadequacies were not economically important as an acute shortage of housing and deep rental subsidies still made them relatively attractive options.

The pattern of SEGs emerging within the public housing sector depends in
part on both the role chosen for public housing (that is, who is eligible) and
the way in which the authority matches households to vacancies. In small
public sectors it could be expected that public neighbourhoods would be

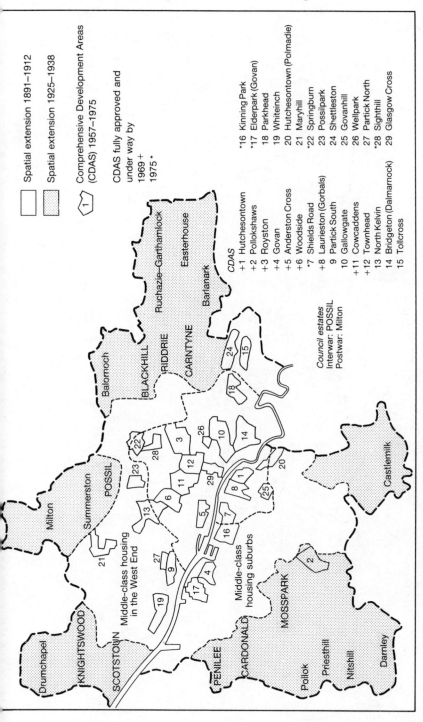

Fig. 6.3. Glasgow Housing and Redevelopment Areas.
Source: Gibb 1983.

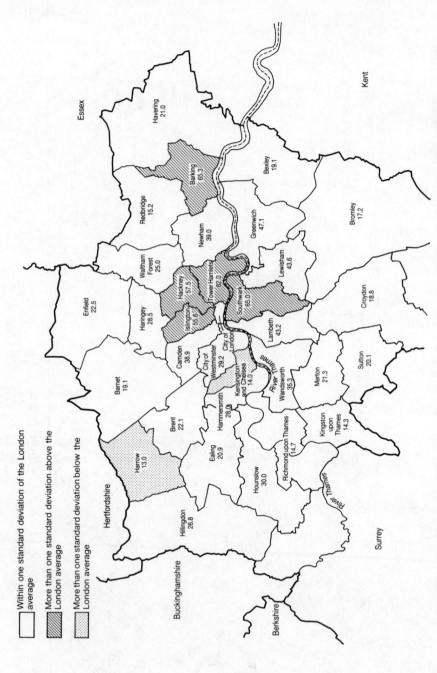

Fig. 6.4. The GLC Area and the London Boroughs: The Percentage of Households in Council Housing, 1981.

Note: The figures shown on the map are percentages. Mean = 30.7. Standard deviation = 17.6.

Table 6.2. *Percentage of Council Tenants in SEGs 1–5 in British Towns and Local Authority Areas, 1981*

Town or local authority	Professionals and managers (SEG 1)	Non-manual workers (SEG 2)	Skilled workers (SEG 3)	Semi-skilled workers (SEG 4)	Unskilled workers (SEG 5)
Hastings	20.6	28.9	31.4	15.0	4.0
Harrow	33.0	30.1	24.7	9.5	2.7
Sutton	29.4	31.2	25.5	10.6	3.2
Redbridge	26.4	31.1	27.7	11.2	3.5
Bury	22.5	22.4	33.8	15.3	5.9
Camden	25.7	34.5	18.4	15.6	5.8
Wandsworth	21.6	32.1	23.8	16.0	6.5
Lambeth	16.9	31.7	23.6	18.6	9.1
Haringey	19.6	29.1	27.3	18.9	5.1
Derby	17.3	20.3	35.9	19.8	6.8
Blackburn	14.0	16.7	39.4	21.3	8.6
Accrington (Hyndburn)	14.0	17.0	40.5	21.9	6.6
Oldham	14.3	17.4	36.5	23.0	8.3
Coventry	15.0	17.7	37.0	24.4	5.9
Wolverhampton	16.5	16.5	40.1	20.6	6.4
West Bromwich	12.4	13.7	43.9	23.3	6.7
Crawley	22.3	23.7	32.9	15.1	6.1
Stevenage	25.1	24.0	32.1	15.0	3.9
Northampton	18.7	21.6	37.8	16.8	5.1
Redditch	19.5	18.7	38.2	19.8	3.7
Plymouth	16.0	24.3	37.1	15.9	6.6
Bath	25.5	29.3	27.9	12.7	4.7
Cardiff	22.3	25.0	31.7	14.4	6.6
Hereford	18.6	22.9	34.2	18.2	6.1
Worcester	19.2	22.5	34.3	19.2	4.8
Newport	17.5	19.2	36.7	17.5	9.0
Chesterfield	14.3	18.0	41.5	19.1	6.4
Sheffield	17.3	17.9	37.9	20.4	6.4
Hartlepool	12.9	14.3	39.8	21.7	11.3
Manchester	13.3	20.5	33.6	22.8	9.8
Total 228 British towns	5.9	13.8	40.3	27.6	12.2
Total in all tenures	20.1	22.2	33.8	17.6	6.2

Source: Isabel Robertson, Town and Regional Planning Department, University of Glasgow (based on Census data).

dominated by lower SEGs, and this expectation is the foundation for present concerns about public sector contraction and 'residualization' (that is, public housing for the poorest only). In fact, Table 6.2 indicates that small public sectors, usually in the South East, have a relatively higher status resident profile (at least in 1981) than those in the North. Such an observation is not implausible in the tight market conditions of the more prosperous parts of

Britain. Thus, in both large and small public sectors, the critical question is the extent to which SEGs are mixed.

There is increasing evidence available to suggest that social, racial, and even religious separation have become commonplace in British public housing (Clapham and Kintrea 1984). The allocation and transfer systems may partly be to blame (systematic evidence does not exist on such matters) but it is also the case that many tenants can now consider private as well as public offers and reject the latter if they do not find them attractive. This implies of course that the 'no choice' tenants must accept offers where they occur, and most frequently they occur in the worst public housing. Whatever the mechanisms involved, public housing in Britain has differentiated in quality as it has matured, and at the same time the worst housing has come to be associated with the lowest income groups. In many British cities it is these neighbourhoods—often no more than two decades old and frequently on the remote edges of the city—that are now the main concentrations of poverty and unemployment. In some areas the loss in socio-economic status of the area reflected the unemployment proneness of existing residents. In others, as implied above, the unattractive nature of these schemes placed them at the foot of the ladder of housing quality within the metropolitan areas and higher status groups have left to be replaced by lower income groups. In Glasgow, for instance, more than half of the continuing households purchasing very low-value properties in older neighbourhoods had most recently resided on a peripheral scheme to which they had been moved in the previous decade. In the Partick area of the city (area 19 in Figure 6.3) almost half the population were moved out to the Drumchapel area in the 1960s. Now a local housing association operating in area 19 has 800 rehabilitated houses and a waiting list of 700 households, more than half of them from Drumchapel, even though association rents are significantly higher than those for local authority dwellings. And, in 1978, more than 55 per cent of council tenants in the city were seeking a transfer to another area. Thus we can conclude that, not only was decentralization partly a publicly induced pattern, but there has been no enduring avoidance of socio-economic segregation and many of the residents would prefer alternative housing.

We can now take public and private developments together and summarize the above sections. The processes of housing decentralization and income separation initiated in the nineteenth century have continued during the postwar period. Rising affluence encouraged the consumption of better housing, thereby continuing the tendency towards decentralization of residences in the private sector, although the conversion of housing from private rental to owner-occupation probably moderated this tendency. Growth in the number of households continued in all of the major conurbations, including London, if the London area is defined to include its Outer Metropolitan Area. Even during the 1970s, the number of households grew in each of the major conurbations, albeit slowly in most. This growth also encouraged decentralization, as did slum clearance in the inner areas and the building of local authority peripheral housing estates in the northern conurbations.

The durability of housing entails that factors influencing residential development in the conurbations over the past 80 years continue to affect the spatial pattern of residential neighbourhoods. Developments in the private market play a large role in some places and in some periods, while in others public sector slum clearance and house-building play a larger role. Because of the processes described above, this legacy of the past, in conjunction with the operation of the private market, tends to concentrate low-income people in more central locations, although there are now signs of change in this pattern. While much attention has been given to the 'gentrification' of housing in inner areas, it has not been a large-scale phenomenon, although it has been important in some neighbourhoods. The spatial pattern of public sector building reinforced this concentration in London, since low-income people have been more likely to be local authority tenants and local authority housing has been concentrated in some inner London boroughs (see Figure 6.4). In other metropolitan areas, large concentrations of local authority housing have developed in suburban areas (for example, Glasgow in Figure 6.3), as well as in inner parts of the metropolitan area.

We now have to consider how such patterns, which were of course influenced by policy, affect local economic development. Before examining specific policies we outline some of the ways in which housing policies can have an impact upon local economies.

Housing–Economy Interactions at the Local Scale

Housing policy, as a collective national activity, is an extremely difficult and diverse entity to identify. Policy may be inaction as well as action by government. In turn, action may consist of controls and statutes and standards as well as financial assistance. And financial assistance may include dwelling-tied subsidies, income-related supports, as well as tax incentives and relief. In order to describe policy impacts, a great deal of precise knowledge about consumer and producer reactions to controls and incentives would be required, even if a unitary national policy existed.

However, a great deal of rehabilitation policy and public sector housing policy is locally structured even if it takes place within tight national financial limits. Thus, in the sections which follow, our empirical material is illustrative rather than descriptive, and we can only hint at the effects of some very major policies and their reform. In reviewing the literature for this chapter we were surprised by the lack of urban-level analysis of housing finance policies, and an examination of such topics internationally indicated that it is only in the USA that such research has been undertaken in any detail. In the rest of the chapter we focus upon financial aspects of policy, although the effects of control are discussed where appropriate.

The previous section stressed how cities consist of neighbourhoods which have some systematic order and, in turn, that lower income groups were most preponderant in less desirable neighbourhoods and that in the private sector this usually implied older, smaller housing. A sharp division between the public and private was observed, both in *modus operandi* and in location. These points are worth reiterating when we recognize that the main elements of housing finance policies are structured by housing tenure (the range of council housing assistance as opposed to tax reliefs), age and condition (the rehabilitation programme), the concentration of poor conditions (Housing Action Areas), and, to an increasing extent, the income of residents (with Housing Benefit increasingly replacing tied council housing subsidies). Assistance with housing costs is not only differentiated by tenure, as there has been no attempt to create a tenure-neutral subsidy system in Britain, but the availability of resources for investment differs sharply across programme headings. Thus the nature and character of neighbourhoods within a British city, in conjunction with local authority policy, are a critical determinant of which policies have an impact upon those neighbourhoods.

In a broad sense, housing finance policy, including tax policy, acts to reduce the costs of living in a particular location or, where new investment or reinvestment decisions are being made, to increase the rate of return to investors, and therefore stimulate investment. In making connections between housing and other urban policies we therefore have to establish the effects of current subsidies—in particular, what other forms of consumption and saving they permit and whether non-portability of subsidies has a particular impact upon household mobility in response to changing economic opportunities. Since detailed household budget studies are not available at the intra-urban scale in Britain, there is little we can say about the local consumption effects of rising house prices or council rents. They may induce more labour-market activity or they may reduce the consumption of imports, or foreign holidays, or the necessities of life. All will have different local economic impacts. However, we can make some comment regarding mobility.

When we consider policies which stimulate housing investment, and particularly reinvestment in older neighbourhoods, we have to examine not only impacts upon internal housing amenity and the immediate neighbourhood but also a wider set of linkage and spillovers. As noted earlier, not only were these linkages neglected throughout the main phases of public sector development, but they still have minimal official recognition in area renovation policy, as for instance in the Green Paper on Renovation Grants (HMSO 1985). In this section, putting investment and current subsidies in a more systematic order, we outline some of the urban economy considerations which flow from the housing sector to the wider urban economy. For brevity we focus on questions about investment and reinvestment spending (again using illustrative examples from Glasgow) and about the impact of housing policies on household mobility.

Possible impacts of housing investment

Current economic research on housing stresses the complexity and durability of housing structures. Thus the phrase 'housing investment', and hence 'rehabilitation', is too generalized as a concept on which to construct a model impact. That is, investment which is focused on the internal features of the dwelling and which is inexpensive to implement may have minimal impacts on neighbourhood regeneration, even if household satisfaction rises dramatically. Rehabilitation policy in the USA, for instance, is often concerned with such small expenditure (Fogarty 1984; Bradbury and Downs 1981), and, where housing authorities in Britain have not pursued active area rehabilitation programmes, 'pepper-potting' of grant aid has a similar small impact. On the other hand, investment may be substantial at the dwelling level with alteration of internal and external features, and the programme may have some perceptible neighbourhood concentration. In the latter instance, a wider set of connections arise. Adjoining neighbourhoods may be affected and there may be stronger linkages with the wider urban economy. The levels of concentration required to promote desirable effects on other parts of the city then become an important question.

The rehabilitation programmes operating in a number of large British cities are a mix of such policies. In Glasgow, for instance, the repair programme financed by Glasgow District Council can result in small-scale, internal, and non-area-targeted investment, but since 1980 targeting of blocks has become the common practice. The Green Paper on Renovation Grants (HMSO 1985) notes that more than forty such schemes currently operate in England, though this seems a rather small number. Where area investment programmes exist, or where rehabilitation-oriented housing associations operate, there clearly is a major programme of substantial housing investment, with internal and external dwelling alterations and with neighbourhood concentrations of investment. The empirical observations noted below are primarily related to the association programme, and the programme initially has a large impact on the lowest-quality private neighbourhoods indicated in Figure 6.1.

A diagrammatic representation of some of the repercussions which can flow from housing reinvestment are indicated in Figure 6.5. For illustrative purposes, consider three routes through the model. In the figure, more immediate and localized effects occur at the bottom left-hand side. First, substantial investment in housing, regardless of structure mix and area targeting, will have a marked effect on the local construction and materials sector of the economy, usually the local metropolitan economy. Where substantial unemployment exists, this represents an unambiguous benefit of rehabilitation spending. In the early 1980s, with an annual rehabilitation programme exceeding £100 million, more than 5000 jobs in Glasgow and the surrounding area were created, primarily in the building trades. This effect could be almost doubled by indirect multiplier (backward and forward) effects. These jobs are not targeted towards residents of lower-income neighbourhoods, but most of the employment beneficiaries are likely to reside in either the public sector or the bottom half of the private market.

Second, substantial internal improvement creates internal resident satisfaction, and all residents in houses in major rehabilitation schemes in Glasgow expressed considerable satisfaction with the outcome and process. There are probably few benefits to non-residents from such spending (see Varady 1981). Substantial improvement in lower-quality structures will either increase public spending or, more probably, squeeze out other forms of housing spending by government, but the impacts will be diffuse and will occur in other cities and regions. In the longer term, in a market system, increased quality may lead to higher income groups displacing poorer residents in the neighbourhood, though this may increase and diversify local service demands. In the housing association programme, with 12 000 completed dwellings in the city, non-price rationing and lettings avoid a direct 'displacement' effect and subsidies largely offset rent increases. If higher income groups do move into a neighbourhood, then policy evaluation will, of course, have to consider questions of social mix and whether or not such residents are moving away from and destabilizing other locations within the broader housing planning area, such as the peripheral housing schemes. Thus, even where rehabilitation focuses on extensive improvement of the internal characteristics of dwellings, there may be complex and long-term changes in the neighbourhood.

The third route through the model is even more complex and, ultimately, elusive to measure. Housing investment will invariably jointly produce environmental change—say, via window-frame pointing, stone-cleaning, and backcourt improvement (all major considerations in the association programme)—and resident interest in 'participation'. If such improvement occurs with sufficient intensity, not only will local residents alter their perception of the neighbourhood but non-local images may also alter. Research indicates in Glasgow that association activity increased the capital values in adjacent neighbourhoods by some 7 per cent more than would have been expected (over the three-year period 1977–80). These spillover benefits are not only beneficial *per se*, but they also raise the value of the local tax base.

Aside from attracting higher-income residents, as in the second route, a revitalized neigbourhood may now enjoy a new status and attract the confidence of individual and institutional investors. If this effect occurs, depending on social mix objectives and views on displacement, government may then contrive a change in the private/public ratio of regeneration finance. In Glasgow, for instance, rehabilitation has helped to shape the context in which new private housing investment occurred in the area in the period 1980–4. New construction sites are usually close to association works. An analysis of the socio-economic characteristics of the purchasers of new private

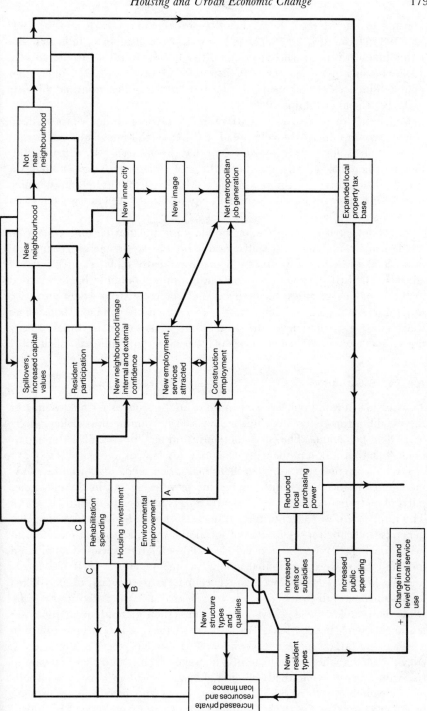

Fig. 6.5. Some of the Repercussions of Housing Reinvestment.

housing in the Glasgow East Area Renewal scheme (GEAR) during 1982
and 1983 (Lamont *et al.* 1985), and their sources of finance, indicates that
individual institutions who would not invest in GEAR in 1977 would do so in
1982. Housing investment more than doubled in this period. Of course, other
policy influences also shaped this outcome, but rehabilitation undoubtedly
had a beneficial attracting effect.

Depending on the scale and structure of the programme, improvement
could become sufficiently widespread that the overall image and confidence
of the city transformed. It is no accident that the '*Glasgow's Miles Better*' cam-
paign, which has greatly improved the city's national image, followed a
decade of intensive rehabilitation work. Municipal public relations, which
may be as important for job creation as city planning, need at least some fac-
tual basis.

In the long term, prospects for mobile job attraction become enhanced,
and housing confidence may spill over into broader social energies promoting
new social and cultural ventures. It is still too early to assess such indirect
effects in the British or Glasgow context, but they should not be disregarded
in the remaking of cities. Such effects must be considered where all major
programmes take place, including the renovation of municipal housing as
well as new construction. In the penultimate section we make broad deduc-
tions about such broad urban effects of other important national policies.

Housing and mobility

In assessing housing policies we cannot restrict our focus to the dwelling or
the neighbourhoods created, but we must also examine how policy affects
household behaviour. The effects of tenure and mobility are likely to be particularly important in influencing economic adjustment.

We have argued that a private housing market tends to concentrate low-
income owner-occupiers and low-income private tenants in the old, inner
parts of metropolitan areas. This clearly affects the social mix of inner areas.
The implication of concentrations of local authority housing for social mix
depends upon the representation of social and income groups in local authority housing. Tables 6.3 and 6.4 show how different socio-economic and
income groups were distributed among the major tenures in 1982. Households headed by someone in a manual occupation, particularly if semi-skilled
or unskilled, are disproportionately represented in local authority housing,
while households headed by someone in a non-manual occupation are
mainly in owner-occupation. Also, if we ignore the first two income groups in
Table 6.4, a large proportion of whom are pensioners, there is a clear tendency for the likelihood of a head of a household being an owner-occupier to
rise with income, while the likelihood of being a local authority tenant falls
with income. This relationship between tenure and income became more
pronounced during the 1970s (Robinson and O'Sullivan 1983). Thus, there

Table 6.3. *Percentage in Tenure by SEG: Great Britain, 1982*

SEG of head	Owner-occupied	Local authority, New Town, or housing association	Private rental
Economically active head:			
Professional	89	2	4
Employees and managers	86	7	4
Intermediate non-manual	77	11	9
Junior non-manual	63	23	9
Skilled manual and own-account non-professional	56	36	5
Semi-skilled manual and personal service	35	50	9
Unskilled manual	27	62	10
Economically inactive heads	43	46	11
Total	56	35	8

Source: 1982 General Household Survey.

Table 6.4. *Percentage in Tenure by Income Group: Great Britain, 1982*

Usual gross weekly income of head of household and spouse (£)	Owner-occupied	Local authority, New Town, or housing association	Private rental
0–30	40	42	17
30.01–40	40	41	18
40.01–60	26	61	13
60.01–80	33	54	11
80.01–100	43	42	12
100.01–120	45	39	10
120.01–140	56	32	9
140.01–160	56	36	5
160.01–200	69	24	4
200.01–250	78	16	3
250.01 or more	90	5	2

Source: 1982 General Household Survey, Table 5.13.

is a clear tendency for areas dominated by local authority housing to have concentrations of economically disadvantaged people.

The association between local authority housing and geographic concentrations of economic disadvantage is strengthened by differences in mobility patterns between tenures. Housing tenure influences both the extent and the direction of household mobility in response to employment opportunities, growth in household income, and changes in family size.

Local authority tenants are more likely to move house than owner-occupiers (Hughes and McCormick 1985), but unless they leave the tenure their movement is constrained by the location of local authority housing and by the difficulty of transferring between local housing authorities. For instance, in London local authority housing is scarce in most of the outer boroughs of

London and beyond, and it is particularly difficult for local authority tenants to move out of inner London. Given the (increasing) tendency for local authority tenants to have low incomes, the operation of local authority housing reinforced the economic segregation produced by the private market in London.

A similar pattern was found in the Manchester area in the late 1970s, as Table 6.5 shows. While local authority tenants had a higher rate of movement than owner-occupiers, most of the movement by inner city council tenants was within the inner city, and local authority tenants were very unlikely to move into the inner city from outside. If we focus on persons (or their partners) responsible for the tenure of their accommodation, then movement into or out of the inner city (Partnership Area) was very unlikely for local authority tenants. In contrast, when owner-occupiers moved, they were very likely to move out of the inner city, probably as part of the process of consuming better housing discussed earlier.

Table 6.5. *Annual Movement in Individual Tenure Groups by Residents of the Manchester–Salford Inner Area, 1978–1979 (Excluding Student Households)*

	Movement rate[a] (%)	Proportion subsequently moving	
		out of inner area[b](%)	into inner area[b] (%)
By tenure sectors[c]			
Owner-occupation (36%)	12	8	3
Local authority (47%)	14	4	2
Private renting			
all (17%)	22	10	5
furnished (2%)	50	30	21
unfurnished (15%)	19	8	4
By individual tenures[c]			
Owner-occupation			
Responsible for tenure (20%)	9	6	3
Other (16%)	16	10	4
Local authority			
Responsible for tenure (26%)	11	2	1
Other (21%)	19	5	3
Private renting			
Responsible for tenure (11%)	19	10	5
Other (6%)	26	11	8

[a] Defined relative to the inner area population within the particular tenure group. In-movement relates to the year preceding the first interview. Out-movement relates to the year between interviews.
[b] 'Inner area' is the Manchester–Salford Partnership Area, which comprises the inner area of both cities.
[c] The percentages in parentheses show the proportion of persons in each group.

Source: Hedges and Prescott-Clarke (1983).

While outward movement by local authority tenants may be less constrained in other conurbations (because of the existence of peripheral local authority housing estates), its direction will still be constrained by the location of local authority housing within the metropolitan area, and by the difficulty of transferring between authorities. Most importantly, local authority tenants find it much more difficult to make long-distance moves than owner-occupiers (Hughes and McCormick 1985), making it harder for them to adjust to changing employment opportunities, which are the dominant influence on long-distance moves. Thus, local authority housing not only attracts people who face greater risks of low income and unemployment, but it also constrains their movement in a way which reduces their economic chances. It is not therefore surprising that analyses of spatial variation in unemployment rates generally find a strong association between a local area's unemployment rate and the proportion of its households in local authority housing (see, for example, Gordon (1985) for the London region).

Private tenants, particularly those in furnished accommodation, are the most mobile (Hughes and McCormick 1985), and this is also illustrated by Table 6.5. Private rental housing is concentrated in old low-quality dwellings in inner areas of cities. For the most part, this is a transitory tenure, used by new households (for example, newly married couples and singles) and long-distance movers into metropolitan areas. These households eventually move into one of the two main tenures. If they move into local authority housing, they often stay in the inner city, but if they enter owner-occupation, they generally upgrade their housing by moving outward. During the latter half of the 1960s, private tenants entering owner-occupation were the largest group of outward movers from inner London, constituting about a third of such moves (Gordon *et al.* 1983). At least in London, entry to owner-occupation was a way for people to improve their housing; they were not forced into owner-occupation as a way of enabling them to move outwards (Gordon *et al.* 1983).

Private rental housing acts as a magnet for young households and newcomers to a metropolitan area (often also young). These people often have low incomes, and they may drive up the local unemployment rate, as Gordon (1985) found in the London region. Areas of concentration of private rental housing may, therefore, be associated with concentrations of low-income people, but the private tenants themselves are often upwardly mobile and geographically mobile. As these young people move up the earnings ladder, they seek better housing by moving away. While the housing may be poor and the tenant's incomes low in an area with a large amount of private rental housing, it may not be cause for concern because the neighbourhood is being used as a stepping-stone by young, mobile households. This may also be the case for old low-quality owner-occupied housing in inner areas. Indeed, areas of poor owner-occupied housing often coincide with areas of private rental housing. This poor housing may also be used as a transitory phase by young owner-occupiers. (There are, of course, many pensioners residing in poor private housing in inner city areas, either as owners or unfurnished tenants. They tend to be immobile because of long-standing ties to the area and because of rent controls in unfurnished tenancies. While their presence may contribute to a concentration of low-income people in inner cities, the focus of this essay is on improving the economic chances of economically active people in inner cities.)

From this perspective, it is local authority tenants who are put at the greatest disadvantage as a consequence of housing policies. They find it very difficult to move out of their local authority area to improve their employment opportunities, and the location of local authority housing, in inner area or peripheral estates, does not generally coincide with the areas in which economic opportunities are more abundant. These difficulties are intensified by the fact that local authority tenants are not drawn randomly from the population, but rather are people who are more likely to have low earnings and to become unemployed. Of course, local authority tenancies may not be the only constraint on their mobility. For instance, Gordon *et al.* (1983) find that less skilled (and lower-paid) workers in inner London rely more on informal or local information sources in searching for housing, such as personal contacts in prospective areas. While this limitation on information constrains their outward mobility, the authors conclude that a shortage of opportunities in local authority housing is the primary reason for their limited outward movement from inner London. Again this in marked contrast to a number of northern conurbations.

The issues we have raised in this section, while they may encompass two particularly important aspects of housing–economy interactions, by no means cover all the important linkages involved. After a brief review of recent changes in housing policy in the next section, we suggest the likely effects arising from the major aspects of current policies in the following section.

A Cameo of Housing Policy 1979–1985

The urban structures discussed in the previous sections largely evolved when government commitments to housing policy spending were growing year by year. By the early 1970s public spending on housing accounted for around one-tenth of the government spending total, or just under 2 per cent of gross national product. This proportion fell after the mid-1970s, partly as a result of attempts at macro-economic control, but also because, as in other countries, there was a growing balance emerging between household numbers and housing units. And at that time the widespread resource needs for rehabilitating the existing housing stock had not yet been recognized. Since 1980, however, government has not viewed sustained cut-backs in housing spending by the state as being pragmatic inconveniences but rather as a redefinition of the appropriate role for government in housing. That is, government's role should be greatly reduced and home-ownership expanded. Clearly, such retrenchment is likely to have urban impacts, not least because overall spending decline has been associated with quite substantial expansion under certain programme headings. Our conclusion is that government policy has had major positive effects in some, but not all, older neighbourhoods— namely, the kind of neighbourhood envisaged to be problematic in the Inner Cities White Paper (HMSO 1977). However, its attempts at dealing with the rapidly emerging and now most difficult run-down public sector neighbourhoods can only be described, even with some exaggeration, as being cosmetic.

The main thrusts in policy spending, of importance here, can be described as follows. Since 1980, current account support for council housing (Housing Support Grant) has been largely removed, and its major concentrations now occur in London and Scotland. Real rents have risen by more than 50 per cent in many authorities since 1980, though more recent increases have been quite low. The effects of these rises have been assuaged, to some extent, for households in receipt of Housing Benefit: thus public aid for housing costs has been largely switched to an individual household income basis. Also of importance, since 1980 local authorities can earn a surplus upon their Housing Revenue Accounts, and this opens up the possibility of using council rental surpluses either to reduce rate charges or to sustain urban public services. At present, however, such surpluses are of modest importance and appear to be a rural or suburban manifestation. Of vital importance in our argument is that public sector investment in council housing is now less than 10 per cent of its 1979 total. Such a downward reduction in spending is consistent with government's spending plans, privatization objectives, and housing 'balance'. It is not consistent, however, with observed neighbourhood deterioration, and indeed abandonment, in the public sector where past inadequate planning, construction, and maintenance policy now leave local authorities facing an estimated bill of £25 billion (more than 10 years' government spending at present levels). Such figures, produced by various pressure groups may be exaggerated, but even if they are 20 cent correct the problem is still serious. In recent years, aside from in Scotland, around three-quarters of housing capital spending has been raised from sales receipts. At present English authorities retain around £5 billion of such receipts on deposit, precluded from spending it by controls on local government expenditure. The coexistence of these accumulated receipts with continuing decline of our poorer neighbourhoods makes neither social nor real economic sense, and it may eventually prove to be bad politics as well. In the following section, we examine structural changes in policy for council housing.

Policies and programmes designed to assist private housing and its rehabilitation have had a quite different experience, at least up until 1984. Council spending to provide mortgages where building societies would not

grant loans has been almost completely removed, but this has had few negative effects so far. For government, in the period 1980–2, have to be
credited, in part, with the changing attitude of building societies towards
housing in decayed neighbourhoods. Societies, with most of the major institutions now having their own internal renewal sections, have desisted from
withdrawing from areas of difficulty, but now, instead, provide initiatives,
often along with local authorites, to promote area rehabilitation. In 1983,
for instance, building societies lent £1.5 billion on advances for housing
repair and improvement—up from £400 million in 1979. Critics could argue
that such shifts merely reflect the relative collapse of new construction coexisting with sustained inflows of funds. However, as the new housing market
recovered somewhat in 1983–4 there was no apparent withdrawal of societies from their new role in such difficult locations as Belfast, Liverpool, and
Glasgow. In relation to private sector rehabilitation grant aid, public expenditure on grant aid for housing improvement and repair (for the UK) rose
from around £100 million in 1978–9 to just over £1000 million in 1983–4.
Not all areas of the country benefited equally, and there is some evidence
that the larger urban centres benefited most. In 1982–3, for instance, the
two cities most cited in this chapter captured around 20 per cent of the
national programme, with Glasgow's share around 7 per cent. In addition
the spending limit for the Housing Corporation increased until 1982–3,
since when it has suffered minimal reductions. Almost half of the corporation's budget—around £650 million in 1983—is devoted to area renewal.
Taking these programmes together, the Conservative government promoted
record spending levels for housing rehabilitation, though curtailment has
now commenced.

The final thrust of government policy, clearly related to the above concerns, was the development of owner-occupation. Since the government's
macro-economic policy reduced inflation, including house-price inflation,
and raised mortgage rates to postwar record real rates while increasing
unemployment, it can be regarded as having an anti-home-ownership stance.
For rising real incomes, secure employment, house-price inflation, and negative real interest rates had all contributed to ownership growth in the 1970s.
However, the negative stance towards the public sector and the very positive
thrust of rehabilitation policy boosted the growth of home-ownership.
Government, in 1980, also promoted a range of 'new initiatives', which have
had a minimal overall impact (of around 20 000 dwellings), although they
may have local neighbourhood significance. Government mainly sustained
high rates of home-ownership growth by introducing, in 1980, the right of
tenants to buy their council house, and to do so with a considerable discount.
Home-ownership growth in the UK, except for Scotland, is now primarily
associated with inter-tenure transfers rather than new construction, and this,
as we note in the next section, has implications for urban form. Finally, and
perhaps of greatest long term significance, the value of tax reliefs on owner-

occupied housing arising from mortgage interest relief has grown steadily and is now approaching £3 billion.

From this very brief description of policy in the 1980s we now turn to a consideration of how these policies, and proposed changes in them, are likely to affect urban areas. We consider, in turn, rehabilitation policy, encouragement of home-ownership, and, finally, the reform of the public sector.

Urban Dimensions of Policy Change

Rehabilitation policy

Although there have been few academic studies of housing rehabilitation policy in Britain, there has perhaps been a prevailing suspicion, following the work of Paris and Blackaby (1979) and Bassett and Short (1981), that such policies have a modest urban impact. They tend, in this view, to impact in a pepper-pot fashion in better neighourhoods. That is, the kinds of economic linkages spelt out in the section on 'housing–economy inter-actions at the local scale' do not come into play. However, in the 1980s, aside from the expansion of improvement grants, policy also now includes large volumes of spending on repairs grants and spending by housing associations: thus a different incidence pattern may prevail. Further, auth-orities and building societies are now more acutely aware of the need to develop area strategies.

The evidence available on the incidence of policy is more convincing than research related to impacts. The recent enquiry into the distribution of grants for England and Wales (HMSO 1985) revealed that the vast bulk of grant aid went to middle- and lower-income households, but generally not to the lowest income groups. The average household income of recipients was about the national average, making it more than double the income of poten-tial programme recipients. The proportions of grants received by the elderly on the one hand and professional/managerial households (potential 'gentri-fiers') on the other were less than their population-based share. First-time home-buyers were a particularly important recipient group. This pattern conforms quite closely to the relatively intense programme pursued in Glas-gow (Maclennan *et al.* 1984). In effect, gentrification exists but does not dominate. Grant aid is directed down market but still often does not have an impact upon the lowest-income households and unfit houses (of which there are still around 1 million in the UK as a whole).

The impacts are less clear. The notion that grant aid increases the values of unimproved property (that is, it is 'capitalized') has mainly anecdotal sup-port, particularly in the South East where the use of grants by developers is important. In Glasgow, some grant capitalization has appeared to exist just below the rateable value limit, but is has not increased in the 1980s, when programmes have expanded, and its upper bound is of the order of 25 per cent of grant values. A more serious concern, which has been at the heart of the recent Green Paper (HMSO 1985), is that a great deal of grant aid is extra-marginal—that is, improvement would have been undertaken anyway. There is no clear evidence on this point.

The large-scale housing improvement programme pursued by housing associations in Glasgow had a major effect upon dwellings and their residents, and at the same time generated important spillovers into surrounding neighbourhoods and the wider metropolitan economy. The environmental improvement programme reinforced such developments and the private sector grants programme had positive, if less apparent, impacts in a range of areas, but particularly those close to the housing association programmes. Table 6.6 uses the framework of the section on housing–economy interactions to provide more detailed information concerning the kinds of impact observed in major rehabilitaton programmes in Glasgow. We would expect similar effects where intensive grant-aid programmes have been pursued.

The government has now proposed in their Green Paper (HMSO 1985) to reform the subsidy system. Loans and grants are now to be means-tested, which although to be welcomed may create problems in multi-unit properties requiring fabric repairs with mixed-income ownership; and grants are now to be directed to only the lowest-income households and worst properties. Since 90 per cent rates of grant aid have not stimulated such action in these areas previously, it is difficult to envisage how the rather vaguely specified new measures will achieve change. The message from the Glasgow experience is to promote neighbourhood 'growth poles' of heavily subsidized housing association activity in such areas, with private finance flowing to surrounding neighbourhoods or association sales taking place in the long term.

Other eligible households, it is now proposed, will receive interest-free equity loans, with individuals repaying on sale a share of their proceeds equal to the initial loan/house-price ratio. Such a measure may well operate without problems in the middle range of property prices, but will be problematic at the cheaper end of the market. There, the combination of high loan/price ratios and the necessity of undertaking non-superficial improvement, with a pronounced 'valuation gap', may trap owner-occupiers into long stays in any house they improve. This may then reduce the mobility of owner-occupiers.

In short, the rehabilitation programmes for older private housing have, where well structured, had major positive impacts on urban construction employment and on urban efficiency, by making quality housing available close to urban cores. New policies proposed are likely to reduce this impact, and the prospects for older neighbourhoods are now less good than under the pre-1985 policy regime.

Table 6.6. *The Impacts of Housing Rehabilitation Programmes in Glasgow*

Policy concern	Housing associations	Improvement grants	Environmental grants
Dwelling stock and quality	Dwelling amalgamation reduces stock by 15%, internal restructuring, full common repair and improvement. Complete removal of quality problems	Adds amenities to adequate houses, not always part of comprehensive tenement revitalization	Stone-cleaning; process creates dust and disorder
Neighbourhood quality	Alters residents' perceptions of the future of the area but leaves most environmental difficulties unresolved	Minimal impact in the East End; cf. other neighbourhoods in the city	Major impact of stone-cleaning and backcourt renewal. Estimated benefits are ten times cost
Rent/cost burden	Rent rise post improvement, even with 90% capital subsidy and Housing Benefit	Grant aid of 75–90% reduces additional mortgage burden but still financing problem for elderly, low income	Grant aided, minimal additional burden
Post-rehabilitation social mix	Minimal change from previous situation as houses are small and allocated administratively. Important long-run issue	No gentrification in East End. Major users are moderate to low income first-time buyers	Most likely instrument to be associated with changing socio-economic status
Resident participation	Major involvement with residents at the key stages of revitalization, growing problem in managing post-improvement stock	None, but possibilities for a concentrated approach to common repair	None, except in housing associations. Need more mechanisms to elicit community view
Neighbourhood economy	Rent rises may be deflationary, possible short-run negative effect on commercial properties. Unclear labour-market effect	Same effect as for housing associations	Minimal in short run, disrupts commercial activity
Spillovers to adjacent neighbourhoods	Housing in boundary zones, raises capital value 10%, attracts private investment and loan finance, less so in GEAR	Not clear, still under investigation	Very substantial, as for housing associations
Metropolitan economy	Major short-run impacts on construction demand for labour and materials. Longer-run effects elusive	Same effect as for housing associations	Same effect as for housing associations

Sources: Housing associations: Maclennan *et al.* (1983); Improvement grants and Environmental grants: Maclennan *et al.* (1984).

Encouragement of home-ownership

Through this essay we have noted how the changing role of building societies, improvement policy, and new patterns of new construction (often encouraged by new initiative schemes) have led to the expansion of owner-occupation in older neighbourhoods. Although this expansion has created some new problems—not least, the rapidly rising level of mortgage defaults and repossessions—it has had major favourable impacts in older neighbourhoods. Households with growing incomes have been retained and more diverse income groups attracted, at least some of them from peripherally located ownership areas. Some new local service demands have been created and, in a broad sense, the policy has expanded location choice for those who wish to be owner-occupiers. Rising real housing costs have also forced first-time buyers down market in the 1980s, and a critical question is whether demand in such areas can be maintained in the longer term. However, since we have already touched upon these issues, we focus here on changes in housing taxation and ownership growth by council house sales.

Reforming owner-occupiers' tax advantages   The exemption of owner-occupiers' net imputed rental income from income taxation constitutes the main source of tax subsidy to owner-occupied housing (Ermisch 1984). It has sometimes been argued that this subsidy encourages suburbanisation by owner-occupiers (Muth 1969), and since they generally have higher incomes than tenants it encourages the concentration of poor people in inner city areas. This is supposed to happen because the subsidy lowers the price of housing, which encourages housing upgrading, the opportunities for which are more abundant and cheaper in suburban areas. However, because the subsidy is proportional to house prices, it reduces the price of housing more in central areas than in suburban ones, which, all else being equal, would discourage suburbanization. The net effect is unclear. Although simulation analysis by Blackley and Follain (1983) suggests that the second effect dominates, this ignores the subsidy's effect on tenure choice. One of the ways the latter occurred was through the shift of housing from private rental to owner-occupation.

The subsidy to owner-occupiers, in conjunction with regulations and controls on rents, meant that a given dwelling would be more valuable if sold into owner-occupation than if rented. This encouraged landlords to sell, which increased the supply of owner-occupied housing, particularly in inner areas where private rental housing was concentrated. Since owner-occupiers tended to be, on average, better off than the former private tenants, who tended to be more heterogeneous in terms of income and social group, this process of conversion from private rental to owner-occupation probably helped retain more of the better-off households in inner areas than would have been the case if the dwellings, which were generally old and of poor quality, had remained in private rental. The owner-occupiers' subsidy provided an incentive to improve this housing for sale, thereby upgrading the housing and the income level of inner city areas. This process may have produced, however, more concentration of the poor at the neighbourhood level, as the poorer people who may have been private tenants moved into local authority housing, which tended to be concentrated in particular neigbourhoods, as discussed above.

This process is, however, probably irreversible. At present, the subsidy encourages entry into owner-occupation by young households, and since house prices are lower further from the centre they are encouraged to move outwards to enter. Removal of the subsidy is unlikely to encourage owner-occupiers to move into local authority housing, but it may encourage young households to do so, thereby slightly reducing outward movement.

With all of these different influences, we cannot really be sure of the effect of the subsidy on location. Its net effect may be small. But it should be recognized that elimination of or reduction in the subsidy could encourage owner-occupiers to leave the inner city, thereby increasing the concentration of poor people there. This also applies to the reduction or elimination of mortgage interest tax relief, which has often been suggested. It would raise the price of housing *to mortgagers* in a manner similar to the effect of taxing net imputed rental income, and it would have similar effects on the location of owner-occupiers with a mortgage.

Council house sales   The 1980 Housing Act gave local authority tenants the right to buy their dwelling at discounts ranging up to 60 per cent, depending on length of tenancy. It appears that somewhere in the neighbourhood of 85 per cent of dwellings sold were houses rather than flats (Sewel *et al.* 1984). They have been mainly semi-detached and terraced houses with gardens, which suggests that only a small proportion of the sales were in inner city areas, or in large peripheral estates, which tend to be dominated by flats. Opportunities for movement within local authority housing have therefore tended to become more spatially concentrated in areas with diminishing economic opportunities, including inner city areas. Thus, tenants' mobility has been constrained further.

It appears that the sales have been to tenants with above-average income. To the extent that these households would have left the area in order to become owner-occupiers, the sales may have reduced the concentration of poor people. There is, however, some evidence that sales have come disproportionately from estates with a 'higher socio-economic profile' (Sewel *et al.* 1984). If this is generally true, the effect of sales on the spatial concentration of poverty will be small. With sales being to better-off households, a large proportion of those who remain face high risks of unemployment and poor earnings opportunities.

Integration of private housing with local authority housing

The processes of housing upgrading by owner-occupiers and entry into owner-occupation encourage movement from inner cities because there are fewer opportunities for upgrading there, and the price of equivalent housing is lower further from the centre. In their study of the London region, Gordon and Lamont (1982) found that the destination of short- to medium-distance migration was related to the distribution of owner-occupied housing opportunities, in particular new private house construction. It might therefore be possible to retain more of the better-off residents in the inner city if more housing for owner-occupation were built there. By expanding public sector housing at the expense of the private sector, some inner city authorities have encouraged owner-occupiers to leave the inner city, thereby abetting the concentration of low-income people in inner areas. The recent experience of London Docklands suggests that there is indeed demand for owner-occupied housing in inner areas. Derelict land deserted by industry could be put to housing use, as was done in Docklands. A better mixture of owner-occupied and local authority housing could improve the social mix of inner areas. The retention of better-off residents in the inner city through the allocation of land for owner-occupied housing could improve the employment opportunities of the less-advantaged by stimulating employment in the service sector as well as reducing the social problems created by the spatial concentraton of disadvantaged people.

Local authority house-building and rents

Another way to improve the economic opportunities for inner city residents in local authority housing would be to encourage the construction of local authority housing in suburban areas of employment growth and to reduce the difficulties of transferring between local authorities. Northern conurbations have, of course, built local authority housing in suburban areas, but they have generally concentrated it in large local authority estates in areas which have not turned out to be areas of expanding employment opportunities. The weakness of the regional economies within which these conurbations lie has limited the effectiveness of a policy of suburbanizing local authority housing. Such a policy may only be effective in the London metropolitan area, the outer parts of which have experienced substantial employment growth, and which is part of a prosperous regional economy.

But, even in the London area, the success of a policy of 'opening up the suburbs' to local authority housing depends on making it easier to transfer between one local authority's housing and another's. This will probably be difficult in a system in which housing is allocated by administrative procedures and by local authorities with different allocation criteria.

Presuming for a moment that the difficulties of transferring between authorities can be overcome, there are, of course, other impacts associated with this policy. Building local authority housing in areas of employment growth would tend to give further impetus to land prices, thereby raising the cost of owner-occupied and local authority housing in these areas. The cost of building local authority housing of a given density would probably still be lower than in inner cities, but owner-occupiers moving into the area would have to pay more while existing owners would receive a capital gain. The economic position of less skilled people already residing in the suburbs would also adversely be affected by the outward movement of local authority tenants because of competition in the labour market.

It is unclear, however, whether the resistance to local authority housing by suburban residents can be overcome. Doing so would be the first step in putting this policy into effect. Its success would then depend on making it easier for tenants to move between local authorities, but even success on this front may do little to reduce the concentration of poor people in inner cities. It could make it worse.

As noted earlier, moves in and around a metropolitan area are affected very little by employment opportunities, even among owner-occupiers. They are mainly made in order to improve a household's housing. Studies by Garner (1980) and English (1979) have suggested that access to good housing in the public sector is highly dependent on the household's ability to wait, which favours households aready in good housing. Suburban local authorities would be very likely to favour households already residing in their area, even if they built more housing and it was open to persons moving into the area. Thus, tenants of inner city local authority housing who were most prepared to wait for an opening in suburban local authorities would be more likely to move outwards. This would favour the inner city tenants with the best housing there, and these tenants tend to have higher income and are less likely to be in the lower SEGs (English 1979). It would therefore be the better-off inner city tenants who would be more likely to move outwards, thereby increasing the concentration of poor people in inner city local authority housing. This tendency would be reinforced by reluctance of the less-skilled to consider movement into unfamiliar areas (Gordon *et al.* 1983). Nevertheless, the labour-market chances of those left behind in the inner city would probably be improved from the supply side because of a tendency, at least in the London region, for persons to decentralize their jobs after decentralizing their residence (Vickerman 1984; Gordon *et al.* 1983).

The tendency for the lower SEGs to get the worst local authority housing clearly entails further concentration of poor people at the neighbourhood level (that is, on the worst estates). This may be associated with additional social problems, although it probably has little effect on their labour-market chances in addition to those effects associated with concentration in the inner city (or peripheral estates).

Clapham and Kintrea (1984) have suggested that persons in the lower

SEGs and with low income are likely to have a strong orientation toward local authority housing, and this stronger orientation makes them more likely to accept a poor-quality dwelling. If, in the proposed spatial widening of opportunities in local authority housing, administrative allocation schemes do not counteract this tendency, then the movement to suburban local authority housing would be socially selective, favouring the upper SEGs, as suggested above. Thus, not only would transfers between local authorities need to be made easier, but allocation schemes would need to be changed to counteract the bias which presently exists in the allocation of local authority housing among social groups.

We have discussed the impact of changing the balance in local authority housing provision between inner areas and the suburbs by encouraging building in the latter, but we have not considered the effect of different aggregate levels of local authority house building. It is important to do so because it has been the present government's policy to reduce local authority house building to the lowest levels experienced since 1945: public sector housing starts during the 1980s have averaged less than half of the previous low. At the same time, the number of households continues to grow, probably at a faster rate than during the 1970s (Ermisch 1985). This means fewer opportunities and less choice in local authority housing, and this is reinforced by council house sales. Entry into owner-occupation is therefore encouraged; and, since it is cheaper to purchase a house in suburban areas, outward movement by the somewhat better-off households is also encouraged, thereby increasing the concentration of poor people in inner city local authority housing.

This process has been further encouraged by a doubling of local authority rents since 1980 (a rise of 60 per cent in real terms). Since this rise does not represent a movement towards rents that reflect the relative size and quality of different dwellings, some tenants have ended up paying more in rent than their dwelling is worth. This encourages them to move into owner-occupation if they can afford it. The least well-off tenants have been sheltered from this rent rise by the Housing Benefit system, and changes in this system proposed in the recent Social Security Green Paper (Department of Health and Social Security 1985) would extend this shelter for the working poor while withdrawing it completely for the moderately poor. The confinement of protection from rent rises to the poorest tenants provides further encouragement for all but the poorest to leave local authority housing.

Conversely, more local authority housing and lower rents would encourage more households to enter the tenure, thereby providing for a better social mix in local authority housing. The ultimate effect on the concentration of economically disadvantaged people would depend on the spatial distribution of local authority house building and the ease of movement by tenants between authorities, as discussed earlier. Of course, more local authority housing offered at rents that do not reflect the relative cost of different dwellings, and allocated by administrative criteria, entails costs in terms of an inefficient allocation of resources. It also means that more households would have their mobility constrained by the difficulty of transferring between authorities. One way of enhancing the mobility of tenants is to move away from administrative allocation criteria towards a market allocation of local authority housing. This takes us into somewhat more radical policy changes (upon which the authors themselves have not yet reached a consensus).

'Freedom of entry' and 'economic rents' in local authority housing   Under this policy, local authorities would set rents on the basis of the demand for and supply of different dwellings in their area. Anyone who could pay the rent could become a tenant, and the authorities would adjust the size and composition of their housing stock in line with rents relative to the costs of construction (see Ermisch (1984) for a fuller discussion). At the same time, poor families would receive a housing allowance which would give them sufficient income to afford housing of a 'decent standard'. If such a policy is to help inner city tenants improve their employment opportunities by moving out, there must be local authority housing to move into in suburban areas. Resistance by suburban residents to local authority housing may again be a barrier. However, if suburban local authorities increase their housing stock in response to rises in rents, then tenants will have an opportunity to move to areas with expanding employment opportunities, provided that they can pay the rent charged in suburban local authority housing. This will depend partly on the standard of housing deemed to be a 'decent standard', upon which the size of the housing allowance would depend.

Even if the standard is set fairly high, there is likely to be a tendency for low-income families to rent cheaper housing, which would tend to be the poorer local authority housing in inner areas. The concentration of poor people in the inner city could indeed be greater than at present because the better-off local authority tenants would probably seize the opportunity to upgrade their housing by moving outwards, leaving the poorer tenants behind. While the housing allowance and the availability of rental housing in the suburbs provide the potential for outward movement by low-income inner city tenants, poor families may place a higher priority on expenditure other than on housing. They may use the housing allowance to reduce the burden of rent on their income and spend little of it on improving their housing. This was indeed the case in the US Housing Allowance Experiments: recipients of allowances did not improve their housing very much, and the allowances did not reduce economic residential segregation significantly (Hanushek and Quigley 1981; Rossi 1981).

Thus, this radical change in the way that local authority housing is priced and allocated, and in the way housing subsidy is paid, would undoubtedly widen the range of housing and employment opportunities open to tenants by removing constraints on their mobility. Movement to areas of better employment opportunities by some local authority tenants would improve the labour-market chances of those left in the inner city by reducing labour supply there, and reduce the differences in unemployment rates between areas. But the concentration of economically disadvantaged persons in inner city areas is not likely to be reduced. It could indeed increase because the policy offers the opportunity of housing upgrading by outward movement to the better-off local authority tenants.

Such a radical shift in rent policy would also require local authorities to earn target rates of return, and hence have accounting, at the estate or neighbourhood level. This could have the beneficial effect of authorities making a calculated assessment of the benefits of reinvestment in decayed housing and in operating higher-maintenance/higher-rent programmes. Such calculations underpin social rental housing renovation in, for instance, Swedish municipal housing companies, or the French Habitation Loyers Moyenne (HLM). In turn, however, this requires that central government relaxes control of local authority housing investment. If government were concerned that normal rates of return on capital were being earned by 'monopolist' local authorities, it could overcome this problem by transferring stock from municipalities to housing co-operatives or housing associations. This could also secure the more decentralized management much praised in the Priority Estates Programme. However, the latter policy alone will not cope with our deteriorating public neighbourhoods. The major obstacles to such changes are likely to be in the Treasury which has such a negative view of the benefits of housing spending (even if unsubsidized) that it is forcing housing investment below the level which could earn a normal rate of return.

What is to be done?

In the light of the strength of the market forces involved, housing policies can only operate at the margin of the problem. They cannot stop decentralization of residences in the owner-occupied sector, but they may be able to slow it down and reduce the concentration of poor people in inner cities, albeit only marginally. The best bet, in terms of both feasibility and outcome, is a set of policies which brings about a better mixture of owner-occupied and rental housing in inner areas. This would diminish the concentration of poor people, thereby breathing some life into areas plagued by what the Inner Cities White Paper (HMSO 1977) called 'collective deprivation'. All of inner city residents would benefit from such an attempt to break down this collective deprivation. Furthermore, the attraction of more affluent residents to an inner area would tend to stimulate the local service economy. Gordon and Lamont (1982) found evidence that strongly suggests that population growth in a local area stimulates growth in service employment, and the effect is particularly strong when the in-migrants are from the more skilled categories. The creation of more jobs by a healthier service sector would improve the labour-market chances for less well-off residents, although this effect is diluted by changes in commuting and the balance of migration caused by the extra employment growth (see Gordon 1985).

One major element in bringing about a better social mix in inner areas (or other run-down neighbourhoods of a city) is housing rehabilitation policy. We have argued that concentrated doses of improvement activity can produce benefits that go well beyond helping the residents of run-down housing. It tends to stimulate private sector housing investment in adjoining areas by improving the physical environment, and the people who move into the new or rehabilitated housing arising from the private investment tend to be owner-occupiers who are better off than many of the existing residents. Thus, the socio-economic mix of the area is improved, which tends to generate the other benefits mentioned above. Furthermore, both the initial and the induced rehabilitation and new building activity spawn construction jobs, many of which go to unemployed and less skilled workers in the city. Various macro-econometric models indeed show that increases in construction activity produce one of the largest effects on the number of manual jobs among all forms of economic reflation.

There is also evidence to suggest that more central parts of urban areas no longer possess a comparative advantage for most industrial activities, including warehousing (Ermisch 1983). Many of the former industrial areas appear more suitable for residential use, and local authorities could encourage them to be put to use for owner-occupied housing. That does not mean that they should stop building local authority housing, but they should try to aim for better balance between the two. Some of the vacant or former industrial land owned by the authorities could be released to builders to build housing for sale at controlled prices, as has been done in London Docklands. While this provides a considerable subsidy (in the form of a capital gain on sale) to the first occupant, this may be money well spent to entice owner-occupiers to these former industrial areas in the inner city. The physical transformation of the area may then make the future attraction of owner-occupiers and other investors to the area possible without subsidy.

The policies which ease movement by local authority tenants to suburban areas appear to be less feasible, and could, on their own, make the inner city problem worse. While they help to decentralize labour supply, thereby improving the labour-market chances of the tenants remaining as well as those moving out to areas of better employment opportunities, the effect on the former group may be reduced by a consequent reduction in local service employment. Since it appears likely that the out-migration by tenants would be socially selective, it would increase the concentration of poverty in inner areas and make it more likely that inner city service employment would suffer.

The present policies encouraging owner-occupation—higher local authority rents, very little local authority building, and continuation of tax subsidies to owners—have similar effects. They 'cream off' the better-off local authority tenants and encourage the better-off young households to enter owner-occupation. These households tend to enter owner-occupation by moving outwards, thereby increasing the concentration of poor people in inner cities.

If, however, policies to help inner city local authority tenants to move out to areas of better employment opportunities were combined with policies to encourage a better housing tenure mixture in inner areas, the policy mix would be even more effective in combating inner city problems. Given the rate of new household formation during the 1980s, local authority housing will need to be expanded to meet the needs of households who are unable or unwilling to become owner-occupiers. The building of new local authority housing could be skewed towards suburban areas, thereby providing new opportunities for movement by tenants from inner areas. At the same time, allocation policies would need to be altered to facilitate this movement. If the local authority allocation and transfer schemes were made less socially selective, then the outward movement by local authority tenants would also reduce poverty concentration.

Housing policies such as these are likely to contribute more to the solution of inner city problems than policies to stimulate economic activity in inner areas by providing low-skilled service jobs and by breaking down the 'collective deprivation' associated with the concentration of poverty. The latter policies are unlikely to have a large effect on unemployment in the inner city because of the induced changes in commuting and migration, and often the types of jobs created do not fit the skills of inner city residents. Attempts to stimulate industrial activity which provide the right type of jobs tend to go against the grain of locational comparative advantage, making it likely that they are not viable in the longer term. Furthermore, these economic policies do little to reduce the concentration of poor people in inner areas. Thus, they are not likely to contribute much to combating 'collective deprivation' in these areas.

Note

1. This chapter is largely based upon research sponsored at the Policy Studies Institute and Glasgow University during the period 1983–5 by the ESRC.

References

Bassett, K. A. and Short, J. R. (1981), 'Housing Policy and the Inner City', *Transactions of the Institute of British Geographers*, New Series, 6, 293–312.
Blackley, D. and Follain, J. R. (1983), 'Inflation, Tax Advantages to Homeowners and the Locational Choices of Households', *Regional Science and Urban Economics*, 13, 505–16.

Boddy, M. J. (1976), 'The Structure of Mortgage Finance: Building Societies under the British Social Formation', *Transactions of the Institute of British Geographers*, New Series, 1, 58–71.

Bradbury, K. and Downs, A. (eds.) (1981), *Do Housing Allowances Work?*, Washington, DC: Brookings Institution.

Clapham, D. and Kintrea, K. (1984), 'Allocation Systems and Housing Choice', *Urban Studies*, 21, 261–70.

Department of Health and Social Security (1985), *Reform of Social Security*, Vol. 1, Cmnd. 9517, London: HMSO.

Diamond, D. B. (1980), 'Income and Residential Location', *Urban Studies*, 17, 1–12.

English, J. (1979), 'Access and Deprivation in Local Authority Housing', in C. Jones (ed.), *Urban Deprivation and the Inner City*, London: Croom Helm.

Ermisch, J. F. (1983), 'Locational Comparative Advantage and Land Use and Employment in London and the South East', ESRC Inner Cities Programme Working Paper, Policy Studies Institute, London.

Ermisch, J. F. (1984), *Housing Finance: Who Gains?*, London: Policy Studies Institute.

Ermisch, J. F. (1985), 'Changing Age Distribution and the Housing Market with Special Reference to Great Britain', in R. D. Lee, W. B. Arthur, and G. Rogers (eds.), *Economic Consequences of Age Distribution in Developed Countries*, Oxford University Press.

Fogarty, T. (1984), 'A Study of Benefits: How to Evaluate Financial Subsidies Implicit in Publicly Aided Rehabilitation for Home-owning', mimeo, University of Pennsylvania.

Gibb, A. (1983), *Glasgow—The Making of a City*, London: Croom Helm.

Garner, C. (1980), 'Residential Mobility in the Local Authority Housing Sector in Edinburgh, 1963–1973', Ph.D. thesis, University of Edinburgh.

Gordon, I. R. (1985), 'Unemployment in London', ESRC Inner Cities Programme Working Paper, Urban and Regional Studies Unit, University of Kent at Canterbury.

Gordon, I. R. and Lamont, D. (1982), 'A Model of Labour Market Inter-dependencies in the London Region', *Environment and Planning A*, 14, 237–64.

Gordon, I. R., Vickerman, R. W., Thomas, A., and Lamont, D. (1983), *Opportunities, Preferences and Constraints on Population Movement in the London Region*, Final Report to the Department of the Environment, Urban and Regional Studies Unit, University of Kent at Canterbury.

Greater London Council (1968), *1966 Annual Abstract of Greater London Statistics*, Vol. 1.

Hanushek, E. and Quigley, J. (1981), 'Consumption Aspects', in K. Bradbury and A. Downs (eds.), *Do Housing Allowances Work?*, Washington, DC: Brookings Institution.

Hedges, B. and Prescott-Clarke, P. (1983), *Migration and the Inner City*, Inner Cities Research Programme, Vol. 9, London: Department of the Environment.

HMSO (1977), *Policy for the Inner Cities*, Cmnd. 6845, London: HMSO.

HMSO (1985), *Home Improvement—A New Approach*, Cmnd. 9513, London: HMSO

Hughes, G. A. and McCormick, B. (1985), 'Migration Intentions in the UK. Which Households want to Migrate and which Succeed?', *Economic Journal*, 95, Suppl., 113–23.

Jones, C. (1978), 'Household Movement, Filtering and Trading up within the Owner-occupied Sector', *Regional Studies*, 12, 551–62.

Jones, C. and Maclennan, D. (1982), *North Sea Oil and the Aberdeen Housing Market*, Report to the SSRC, London.

Kain, J. and Quigley, J. (1975), *Housing Markets and Racial Discrimination*, New York: National Bureau of Economic Research.

King, A. J. (1976), 'The Demand for Housing', in N. Terleckj (ed.), *Household Production and Consumption*, New York: National Bureau of Economic Research.

Lamont, D., Maclennan, D., and Munro, M. (1985), 'New Private Housing in the East End of Glasgow', mimeo, Centre for Housing Research, University of Glasgow.

Maclennan, D. (1981), 'Tolerable Survival and the Central Cities', in M. Gasken (ed.), *The Political Economy of Tolerable Survival*, London: Croom Helm.

Maclennan, D. (1982), *Housing Economics: An Applied Approach*, London: Longman.

Maclennan, D. (1985), 'Individual Patterns of Housing Expenditure: An Analysis for Glasgow', mimeo, University of Glasgow.

Maclennan, D., Brailey, M., and Lawrie, N. (1983), *The Activities and Effectiveness of Housing Associations in Scotland*, Edinburgh: Scottish Office.

Maclennan, D. and Jones, C. A. (1984), 'Credit Rationing and its Area Impact', mimeo, University of Glasgow.

Maclennan, D., Robertson, D., Munro, M., and Carruthers, D. (1984), *The Glasgow Database Project*, Report to Glasgow District Council.

Muth, R. (1969), *Cities and Housing*, University of Chicago Press.

Paris, C. and Blackaby, R. (1979), *Not Much Improvement*, London: Heinemann.

Pollakowski, H. (1982), *A Model of Residential Location Choice*, New York: Ballinger.

Richardson, H. W. and Aldcroft, D. H. (1968), *Building in the British Economy between the Wars*, London: George Allen and Unwin.

Robinson, R. and O'Sullivan, A. (1983), 'Housing Tenure Polarisation', *Housing Review*, July–Aug., 116–17.

Rossi, P. H. (1981), 'Residential Mobility', in K. Bradbury and A. Downs (eds.), *Do Housing Allowances Work?*, Washington, DC: Brookings Institution.

Segal, D. (ed.) (1980), *Neighbourhood Economics*, New York: Academic Press.

Sewel, J., Twine, F., and Williams, N. (1984), 'The Sale of Council Houses: Some Empirical Evidence', *Urban Studies*, 21, 439–50.

Varady, K. (1981), 'The Spillover Effects from Environmental Factors', paper presented at a meeting of the Regional Science Association (Southern Branch), USA.

Vickerman, R. W. (1984), 'Urban and Regional Change, Migration and Commuting—The Dynamics of Workplace, Residence and Transport Choice', *Urban Studies*, 21, 15–30.

Young, K. and Garside, P. (1982), *Metropolitan London: Politics and Urban Change, 1837–1981*, London: Edward Arnold.

Wilkinson, R. K. (1973), 'The Income Elasticity of the Demand for Housing', *Oxford Economic Papers*, 25, 361–77.

INDEX

Aberdeen 13, 17, 49, 55–7 *passim*, 61, 68, 69, 71, 72, 74, 168, 169
Adult Training Strategy (ATS) *see* training
advice 134, 149
age factors 5, 77, 79, 91–4 *passim*, 96, 98–102, 107–9 *passim*, 183–4
agriculture 46, 47
Airdrie-Coatbridge 58–60 *passim*, 68, 72, 73, 75, 76
Aldcroft, D. H. 170
Aldershot 13, 55–7 *passim*, 61, 68–70 *passim*, 72, 73, 75, 76
allowances
    housing 195
    training 128, 130, 131, 139, 140, 148, 150 *see also* Enterprise∼Scheme
amenities 9, 33–4, 161–3, 170
Amsterdam 41
apprenticeships 108, 119, 125, 130, 131, 138
Arnhem 41
Ashton and Hyde 13

Baden-Württemburg 17, 24
Barnsley 68–70, 72–4 *passim*, 76
Barnstaple 150
Basildon 49, 54–7 *passim*, 61, 68, 69, 71, 72, 74–6 *passim*
Basingstoke 54–7 *passim*, 61, 69, 71, 72, 74, 76
Bassett, K. A., 187
Bath 50, 69, 70, 75, 76
Bavaria 17, 24
Bedford 55–7 *passim*, 61, 68, 71, 73, 75, 87
Begg, Iain 4, 31, 44–77
Belfast 186
Berg, L. van den 9
Berkshire 85, 86
Birch, D. 33
Birkenhead and Wallasey 13, 45, 56, 58–60 *passim*
Birmingham 3, 12, 13, 50, 56, 58–60 *passim*, 70, 71, 74, 75, 119, 140, 149, 168
Blackaby, R. 187
Blackburn 58–60, 68–70 *passim*, 73
Blackley, D. 190
Blackpool 13, 17, 50, 68, 72–4 *passim*
Boddy, M. 78
Boddy, N. J. 168
Bolton 13
Bournemouth 13, 68, 73–5 *passim*
Bradbury, K. 177
Bradford 13

Bremen 17
Brenner, M. H., 120
Brighton 13, 69–74 *passim*, 76
Bristol 3, 5, 12, 13, 72, 73, 78
brown-field sites 168, 169
Browning, H. C. 46
Buck, Nick 5, 77–115, 120
Buckinghamshire 85, 86
building societies 7, 185–7 *passim*, 190
Burnley 68, 70–2, 74–6 *passim*
Burridge, P. 89
Burton-upon-Trent 68, 70, 72–6 *passim*
business services 42, 50, 51; studies 142

Cambridge 13, 50, 55–7 *passim*, 61, 69, 70, 73, 75, 83
Cannock 68, 69, 71–6 *passim*
Cardiff 13, 58–60 *passim*, 71, 72, 75, 76, 150
Carleton, D. W. 32
Carlisle 68, 70–2 *passim*, 76
Charlton Training Centre 139
Chatham 69, 70, 72, 75, 76
Chelmsford 69–72 *passim*, 75
Chester 50, 70–2, 74–6 *passim*
Chesterfield 69, 71, 73, 74, 76
Chester-le-Street 68, 69, 71–3 *passim*, 75, 76
Cheltenham 55–7 *passim*, 61, 69, 72–6 *passim*
Cheshire, P. 10
Clapham, D. 174, 193–4
class factors 77, 90, 91, 103, 109 *see also* socio-economic factors
clerical work 32–3, 84, 86, 123
Cleveland 144
Clydeside 98, 106
coal industry 44, 49
Colchester 55–7 *passim*, 61, 68, 70, 71, 74–6 *passim*
community business 117, 141, 147, 152
    Strathclyde Community Business Ltd 141
Community Enterprise Programme 135
Community Programme 122–4, 128–31, 135–7 *passim*, 143, 146–7, 151, 152, 155, 156
Community Project Agencies 130
Community Projects Foundation 134, 148
commuting 40, 42, 87–91 *passim*, 112, 163, 168, 197, 198
competition 4, 47–8, 107
    for jobs 5, 78, 141, 161, 193

concentration, population 2, 5, 7, 112, 160–2 *passim*, 175, 177, 180, 181, 184, 190–8 *passim*
construction industry 7, 47, 83, 107, 123, 130–1, 168, 170, 177, 186, 188, 190, 192–5, 197, 198
consumer-oriented sector 21, 22, 36, 50, 51, 81
Coombes, M. G. 40
costs 30, 32–3, 81, 124–5
counter-urbanization 80–1, 83 *see also* migration
Cousins, M. J. 103
Coventry and Nuneaton 13, 50, 58–60 *passim*, 68, 71, 72, 74–6 *passim*
Crawley 45, 54–7 *passim*, 61, 69–73 *passim*
crime 33–4, 161
Curran, M. 103

Dabson, B. 155
Darlington 68, 71, 72
Dearne Valley 45, 49, 58–60 *passim*, 68, 69, 72–4 *passim*, 76
decentralization 7, 78, 80–1, 83, 105, 165, 167–70 *passim*, 174, 193, 196
defence industry 4, 36, 38, 42
Denham Court Training Centre 139–40
deprivation 5, 7, 161–2, 196, 198
Derby 13, 25, 29, 72–5 *passim*
Detroit 21
Devon 83, 150
Diamond, D. B. 167
disabled 99–102 *passim*, 113 n.7, 131, 134, 139, 144
distributive services 20
Dixon, R. J. 105
Docklands 169, 192, 197
Doncaster 69, 70, 72–4 *passim*, 76
Donnison, D. 108
Dorset 83, 86
Downs, A. 177
Dumfries 150
Dundee 13, 70, 71; Training for Employment and Enterprise Project 148

'Earn and Learn' scheme 143
earnings 5, 10, 89, 90, 110, 150, 184
East 3, 15, 16, 36, 82
East Anglia 15, 81–3 *passim*, 86, 87, 97, 106
Eastbourne 55–7 *passim*, 61, 68–71 *passim*, 74, 75
Edinburgh 13, 17, 50, 70, 71, 73–6 *passim*
education 4, 21, 22, 31, 36–8 *passim*, 50, 55, 91, 108–9, 141–3
    Merseyside~Training and Enterprise Ltd (METEL) 143, 152
    non-advanced further (NAFE) 121, 122, 125, 157 n.8

Technical and Vocational~Initiative (TVEI) 119, 121, 122, 125, 145, 157 n.8
employers 6, 107, 108, 119, 120, 124, 126–31 *passim*, 133, 137, 138, 144, 153, 155
employment *see also* manufacturing; services
    change 11, 24, 26–9 *passim*, 34, 37, 39 n.1, 41, 51, 59, 62, 80–91 *passim*, 97, 103–5 *passim*, 110, 111
    conditions 85, 99, 111
    creation/growth 1–7 *passim*, 10, 12, 15, 19–21, 22, 25–8, 34–6, 38, 49, 54–6, 59, 60, 62–3, 77–115, 120–3, 144, 177–8, 196
    exchanges 45–6
    female 54, 81, 83–6 *passim, see also* women
    opportunities 2, 5, 78, 122–5, 135, 138–41, 182–4, 192, 193, 195–7 *passim*
    preservation 5, 112
    self- 116, 122, 124–5, 145–50
    stability of 78, 80, 85, 103–4, 107–10 *passim*
    'sub-' 104, 105, 112
Employment and Training Act (1973) 119, 133
Employment, Department of 44, 121, 124, 125, 132
engineering 47, 119, 138
English, J. 193
Enterprise
    Allowance Scheme (EAS) 116, 121, 122, 124–5, 147, 148, 151, 152
    Community~Programme 135
    Dundee Training for Employment and~Project 148
    Graduate Programme 145
    Training for~Programme (TEP) 145–6, 157 n.12
    trusts 137
environmental factors 7, 33–4, 49, 160–2, 167–8, 178
Ermisch, John 7, 160–200
estates, industrial 129
ethnic factors 5, 77, 79, 92–6, 99–102, 107–9 *passim*, 113 n.7, 120, 134, 135, 138–40 *passim*, 142
European Economic Community 82, 116, 132, 136, 153, 154
European Social Fund (ESF) 132, 136, 137, 139, 141, 143, 144, 148, 151, 152
Exeter 13, 55–7 *passim*, 61, 68–71, 73–6 *passim*, 150

Falkirk 69, 71–3 *passim*, 75, 76
Fife, South East 69, 71, 72, 76
financial services 42, 107, 118
Fogarty, T. 177
Follain, J. R. 190
Forbes, J. F. 12

Fothergill, S. 17, 23, 32, 33, 81
Foundation for Arts and Crafts Enterprise 149–50
France 196

Garner, C. 193
Garside, P. 170
Gatwick Airport 85
gender factors 54, 91, 99
Genesis Project 149
gentrification 169, 175, 187
Germany, West 2–4, 10–12, 14–21, 23–7 *passim*, 29, 34–8, 40–2 *passim*, 119, 125
Gershuny, J. I. 46
Glasgow 3, 12, 39 n.4, 45, 56, 58–60 *passim*, 118, 119, 125, 129, 137, 144, 146, 147, 163, 164, 166, 168–71 *passim*, 174, 175, 177–8, 180, 186–9 *passim*
    East Area Renewal scheme (GEAR) 180
Gloucester 68, 70, 72, 76
Goodman, J. F. B. 120
Gordon, Ian 5, 28, 77–115, 165, 167, 183, 184, 192, 193, 196, 197
government services 4, 20–2 *passim*, 36, 38, 42, 48, 55
grants 128, 130, 131, 134, 187–8
    Employment~Scheme, Training and (TEGS) 144
    Housing Support 185
    improvement 187–8
    recruitment 144–5
    renovation 176, 177, 186, 187
Gravelle, H. 120
Greaves, K. 110
Greenock 58–60 *passim*, 68, 69, 71–3 *passim*
Grimsby 68, 69, 72–6 *passim*
groups, industrial 47–8, 64–7
growth, economic 3, 4, 6, 12, 107
Gudgin, G. 17, 23, 46, 81
Guildford 13, 72, 73

Hackney 141–2
Halifax 68, 70, 74
Hall, P. 9, 23, 80
Hamburg 17
Hamilton and East Kilbride 69, 72–6 *passim*
Hampshire 82, 86
Hanushek, E. 195
Harlow and Bishop's Stortford 45, 68, 71–4 *passim*
Harrogate 55–7 *passim*, 61, 68, 69, 72–6 *passim*
Hartlepool 58–60 *passim*, 68, 69, 71–4 *passim*, 76
Hastings 70, 74–6 *passim*
Hatfield 45
Hausner, Victor 1–8
Hay, D. 9, 23

Health and Social Security, Department of (DHSS) 150, 194
health services 55
Health Services and Public Health Act (1968) 134
Heathrow–Gatwick area 51–3
Hedges, B. 167
Hemel Hempstead 68, 69, 72, 76
Hertford 55–7 *passim*, 61, 68, 72, 73, 76
High Wycombe 13, 25, 29, 55–7 *passim*, 61, 69–71 *passim*, 73, 75
horticulture 123
hotels/catering 138
housing 5, 7, 81, 102–3, 112, 131, 155, 160–200
    Act (1980) 191
    Action Areas 176
    associations 178, 187, 188, 196
    Benefit 176, 185, 194
    Centre for~Research 169
    Corporation 186
    owner-occupied 165, 167, 180–4 *passim*, 186, 188
    peripheral estates 125, 170, 174, 175, 178, 183, 184, 191, 192
    policy 2, 7, 175, 184–98
    private 7, 81, 103, 168–70, 180, 183, 184, 192
    public 7, 99–101, 103, 170–84 *passim*, 191, 192–5, 197, 198; sales of 7, 185, 186, 191, 194
    Revenue Accounts 185
    subsidies 7, 176, 187–8, 190–1, 195, 197
    Support Grant 185
    tenure 100–3, 176, 181–3
Huddersfield 13, 58–60 *passim*, 68, 70, 71, 74, 76
Hughes, G. A. 181, 183
Hull 13, 68, 74
Huntingdonshire 85, 86

image 178, 180
income elasticity 167–8
industrial-military sector 21, 22, 36, 38
information technology 118; Centres (ITeCs) 130, 149
infrastructure 129
Inner Area Partnership/Programmes 135
Inner Urban Areas Act (1978) 133
innovation 6, 7, 137–56
insurance 42, 118
investment 7, 32, 81, 119, 129, 176–80, 185, 196, 197
Ipswich 13, 68, 71, 73, 83
iron industry 44, 54, 56

Japan 119
Job Release Scheme 122

Jobcentres 121
Jones, C. 165, 168

Kain, J. 167
Kettering 58–60 *passim*, 69, 70, 72, 74–6 *passim*
King, A. J. 167
Kintrea, K. 174, 193–4

labour force 6, 9, 11, 30–1, 37, 38, 82, 118
    administrative, professional, technical
        (APT) 83–6 *passim*, 94, 95, 107
    assembly 97
    black 92, 94, 95, 99–102, 107–11 *passim*,
        113 n.7, 142
    female *see* women
    manual 83, 84, 86, 90, 92, 94, 95, 104–7, 111
    skilled 30–1, 33, 111; semi- 31, 43, 90, 92,
        94, 99, 102
    unskilled 31, 33, 43, 78, 90, 92, 94, 99,
        104–7, 110, 111, 123
labour markets 1–2, 5, 7, 10, 32, 41, 77–115,
    118–22, 126–8, 151, 153–6, 161
Lamont, D. 169, 180, 192, 196
Lancashire 131
land 30, 33, 81, 129, 168, 169, 192, 193, 197
Leeds 130
legal factors 133–4
Leicester 69–71 *passim*, 75
Lewisham 150
Lincoln 68, 71, 75
linkage effects 28
Liverpool 13, 45, 56, 58–60 *passim*, 135, 143,
    148, 152, 186
local authorities 6, 116, 117, 119–20, 123–4,
    127, 129–34, 153, 170–86, 191–8 *passim*
Local Collaborative Projects 155
Local Government Act (1963) 134; (1972)
    133, 134
London 12, 28, 45, 51–4 *passim*, 68, 73, 76, 78,
    79, 81–3 *passim*, 85, 90–2 *passim*, 94, 95,
    97–9, 101–7 *passim*, 109, 112, 120, 126,
    130–2 *passim*, 136, 138, 139, 165, 167, 169,
    170, 172, 174, 175, 181–5 *passim*, 192
    County Council (General Powers) Act
        (1947) 134
    Greater~Council (GLC) 130, 131, 133–4,
        136, 138, 139, 151
    Greater~Training Board (GLTB) 131–4
        passim, 139, 140
Lower Saxony 17
Luton 50, 68, 69, 71, 73–5 *passim*

Maclennan, Duncan 7, 160–200
Maidstone 55–7 *passim*, 61, 68, 71, 72, 75
maintenance work 33, 138
Manchester 70, 165, 167, 182
manpower policy 2, 6–7, 116–59; national 117,
    121–9, 152–4 *passim*

Area~Boards 155–6
Manpower Services Commission (MSC) 6, 7,
    116–19, 121–3, 127–40 *passim*, 142,
    145–56 *passim*
Mansfield 68–76 *passim*
manufacturing 3–4, 22–30, 37, 47–50, 83
    employment in 3, 4, 17, 23–30, 36–7, 38,
        46–54, 107
marital status 77, 99–101, 107–9 *passim*
markets 47, 49, 51, 149
Mason, C. 120
McArthur, Andrew A. 6, 116–59
McCormick, B. 103, 181, 183
McGregor, Alan 6, 116–59
Medway Towns 3, 13, 17, 150
Merseyside 104, 106, 119, 124, 127–9, 135–6,
    155
    ~Education, Training and Enterprise Ltd
        (METEL) 143, 152
metropolitan fringe 83, 85, 87, 95
Middlesborough 13
Middleton, A. 129
Midlands 56, 101, 109, 112, 123, 126
    East 15, 82, 106
    West 82, 106, 119, 120, 126, 128, 129, 131,
        132
migration 28, 87–91 *passim*, 104–5, 198 *see also*
    mobility
    in 11, 86, 87, 89, 94, 111, 196
    out 4, 5, 7, 11, 77, 111, 112, 165, 167, 183,
        191, 193–5 *passim*, 197, 198;
    international 104, 105
Miles, I. D. 46
Milton Keynes 54–7 *passim*, 61, 68, 70, 74, 75,
    82, 87
mining 21, 22, 47, 54, 83
Miscellaneous Provisions Act (1982) 133
mobility 2, 4, 5, 7, 28, 78, 79, 87–91, 104–7,
    110–12 *passim*, 160, 162, 165, 167, 176,
    180–4, 191–5 *passim*, 197, 198
Molho, I. I. 89, 113
Moore, Barry 4, 44–76
Morecambe 68, 69, 72–5 *passim*
mortgages 102–3, 191
Motherwell and Wishaw 13, 58–60, 68–70
    passim, 73, 74, 76
Muth, R. 190

National Economic Development Council
    (NEDC) 119
Neighbourhood Energy Action Project 135
Netherlands 2, 10, 14, 23–4, 26, 34–6 *passim*,
    41
new firms, generation of 30, 32, 145–6
New Towns 4, 5, 50–4 *passim*, 60, 62, 64, 82–3,
    85–7, 89, 92, 94, 95, 103, 110, 111
Newcastle 3, 12, 13, 168
Newport (Wales) 58–60 *passim*, 69, 72

Norris, G. M. 101, 103
North 4, 5, 7, 56, 62, 64, 81, 82, 86, 94, 95,
    101, 104–7 *passim*, 109, 111, 112, 118, 123,
    126, 173, 184, 192
North Rhine-Westphalia 17
North West 62, 82, 104, 106, 120, 123, 125,
    126
Northampton 13, 25, 29, 55–7 *passim*, 61, 69,
    71, 73, 82
Norwich 13, 71, 73, 76, 83
Nottingham 13, 25, 68, 70
Noyelle, T. J. 1, 4, 20, 21–2, 46

Oldham 13
O'Sullivan, A. 180
Owen, D. W. 86
owner-occupiers 7, 103, 112, 165, 167, 180–4
    passim, 186, 188, 190–4, 196–8 *passim*
Oxford 13, 50, 69–71 *passim*, 75
Oxfordshire 85, 86

painting 123, 131
Paris, C. 187
pay 28, 32 *see also* wages
performance, economic 9–43
peripherality 4, 59, 60, 62 *see also* housing,
    peripheral estates
Peterborough 54–7 *passim*, 61, 68, 70–2 *passim*,
    75, 82, 87, 150
Pilgrim Trust 121
planning 5, 153–6
Plymouth 69, 71, 74, 76, 83
Policy for the Inner Cities 160–2, 185, 196
political factors 130
Pollakowski, H. 167
polytechnics 142
Pontypridd 13
population change/growth 9, 44, 82, 86, 94,
    95, 97, 110, 163, 167
Portsmouth 68, 70, 72, 76
Potteries 58–60 *passim*, 69, 73–6 *passim*
poverty 2, 160, 161, 174, 197
Prescott-Clarke, P. 167
Preston 71, 72
prices, house 165–7, 176, 190, 193
Priority Estates Programme 196
private sector 6, 110, 134, 137, 140, 148, 153
    see also employers
producer services 4, 20, 48
professionals 31, 33, 43, 48, 83–4, 90, 107
profit gradient 32
Project Fullemploy 134, 140, 149
promotion 129, 180
public sector 38, 50, 51, 85, 107, 109, 170–84

qualifications 79, 91, 98–102, 107–10 *passim*
Quigley, J. 167, 195
quotients, location 21–2, 54–7, 59, 60, 64

Reading 55, 69, 72, 73, 75, 76, 83, 87
recruitment 111–12, 131–2, 144–5
regional factors 3, 15–17, 23–4, 27, 29, 36, 37,
    104–5, 111, 112, 126–7
    policy 62, 81, 112, 119
rehabilitation/renovation, housing 131, 168,
    169, 175–80, 186–9, 197
rents 7, 174, 176, 184, 185, 190, 194–6 *passim*
research institutes 30
residential preferences 44, 49, 81, 88 *see also*
    housing
retirement, early 122
Rhineland 17
Richardson, H. W. 170
Robinson, R. 180
Rossi, P. H. 195
Runcorn 68, 69
rural areas 17, 83, 85, 87, 95

Samuel, P. J. 120
Sandwell 149
Schleswig-Holstein 17
Scotland 56, 62, 82, 104, 106, 123, 125, 141,
    185, 186
Scottish Development Agency (SDA) 144,
    147, 148
Scunthorpe 49, 56, 58, 68, 71–3 *passim*, 75,
    76, 79
Segal, D. 167
service sector 1, 3, 4, 20–2 *passim*, 37, 42, 44,
    46, 50, 51, 57, 64, 102, 118, 196
    employment in 7, 25–9, 36–7, 38, 47–54, 84,
    85, 86, 97, 102, 107, 118, 192, 196–8
    passim
Severnside 82
Sewel, J. 191
Sheffield 58–60 *passim*, 69, 72, 131, 138–9, 150
shipbuilding 44, 54
Short, J. R. 187
Singlemann, J. 46
size, city/urban area 4, 9, 17–21, 36, 51–63
    gradient 18, 19, 36
skills 5, 6, 30–1, 33, 37, 38, 42–3, 77, 118, 120,
    127–8, 138, 141–3, 149–50, 198
    Centres 127–8, 139, 157 n.11
Slough 68, 71, 73, 75
slum clearance 170, 175
small firms 144–6
socio-economic factors 7, 31, 100–1, 162–75,
    181, 191, 193, 194, 197
Somerset 83, 149
South 3, 4, 5, 7, 15–17 *passim*, 32, 36, 44, 49,
    50, 54, 64, 81, 82, 86, 94, 95, 98, 105, 107,
    109, 111, 140
South East 15, 81, 82, 87, 97, 106, 112, 118,
    120, 123, 126, 128, 173, 187
South West 15, 81–3 *passim*, 87, 97, 106, 123,
    126

Southampton 13, 75
Southend 55–7 *passim*, 61, 69, 73, 74, 76
Southport 69, 71, 74, 75
spatial factors 81, 162–3, 165, 167
specialization 4, 20–2 *passim*, 30, 36, 38, 47, 48–56, 63–4
Spence, N. A. 80
St. Albans 68, 69, 72
St. Helens 68, 70, 73–6 *passim*
Stanback, T. M. 1, 4, 20–2 *passim*, 46
steel industry 44, 49, 55, 56
Stoke-on-Trent 13
Strathclyde 141–4 *passim*
structural change 1, 4, 7, 46–7, 118
structure, industrial 9, 23–5, 29, 30, 36–7, 58, 59, 64
subsidies
    employment 121–2, 124–5, 144–5
    housing 176, 187–8, 190–1, 195, 197
suburbs 165, 167, 170, 175, 190, 192–5 *passim*, 198
Sunderland 13, 58–60 *passim*, 68, 73, 74
Sussex, West 85, 86
Swansea 72, 75, 76
Sweden 196
Swindon 69, 71–4 *passim*

Tamworth 71–6 *passim*
taxation 176, 186–7, 190–1, 197
Technical and Vocational Education
    Initiative (TVEI) 119, 121, 122, 125, 145, 157 n.8
technology 38, 44, 50, 51, 55, 59 *see also* information
Teesside 68, 69, 71, 72
Telford 68, 71, 72, 74–6 *passim*, 83, 87
Temporary Short Time Working
    Compensation Scheme (TSTWCS) 121–2, 124
textiles 44
Thames Estuary 58–60 *passim*, 73
Thanet 68, 74, 75
Thirlwall, A. P. 105
Torbay 68–71 *passim*, 74
Torquay 150
Trade and Industry, Department of 118, 119
trade unions 124, 130, 133
training 2, 5, 6, 38, 85, 116, 119, 125–56
    adult 127–9, 131
    Adult Training Strategy (ATS) 116, 119, 128, 145, 153
    Apprentice~Scheme 138
    Charlton~Centre 139
    Denham Court~Centre 139–40
    Dundee~for Employment and Enterprise Project 148
    Greater London~Board 131–4 *passim*, 139, 140
    Industrial~Boards 137 nn. 2, 7

Merseyside Education~and Enterprise Ltd (METEL) 143, 152
~and Employment Grant Scheme (TEGS) 144
~for Enterprise Programme (TEP) 145–6, 157 n.12
'~on Tap' 143
~Opportunities Scheme (TOPS) 127–8
vocational 38, 119, 138–42
Workshops 130, 135, 149
youth 125–7; Youth Training Scheme (YTS) 116, 119, 121, 125–7, 129–31 *passim*, 137, 139, 140, 149–50, 155, 156
transport industry 83, 123
turnover, employment 102, 107
Tyler, P. 32
Tyneside 56, 58–60 *passim*, 104, 106

unemployment 2–5, 10–12 *passim*, 14, 15, 19–22, 24–7, 34–6 *passim*, 38, 41, 42, 77–9, 86–112, 116–25, 128, 153, 174, 183, 196, 198
    black 94, 95
    British~Research Network 134
    duration of 94–6, 97, 102, 110, 117, 120
    female 95, 98, 99
    long-term 6, 95, 117, 120–2, 143, 147, 152, 153
    structural 118
    ~Alliance 134
    youth 94, 95, 98, 116, 148
United States 1–4 *passim*, 10, 12–29, 31–8, 40–2 *passim*, 46, 119, 151, 161, 167, 175, 177, 195
    Midwest 17–19 *passim*, 23–4, 27, 29, 31, 32, 34, 36
    Northeast 17–19 *passim*, 23–4, 27, 29, 31, 32, 34, 26
    South 17
    West 17, 36
universities 30, 142
Urban Aid 135, 151, 152
urban area definition 40–1
Urban Programme 135, 138
utilities, public 48

value added 10
Varady, K. 178
vehicle industry 21, 47, 50, 51, 138
Vickerman, R. W. 193
Voluntary Projects Programme 148–9, 153
voluntary sector 116, 117, 123–4, 132, 134–7, 148–9, 153
    National Council of Voluntary
        Organizations (NCVO) 132, 134, 135
wages 9, 28, 32–3, 97, 130, 144
Wakefield and Dewsbury 70, 72, 73, 75

Wales 62, 82, 104, 106, 123, 126
Walsall 13, 150
warehousing 197
Warr, P. B. 120
Warrington 50, 68, 72, 73, 76, 83
Wichita 17
Widnes 68, 69, 75, 76
Wigan 13
Wilkinson, R. K. 167
Wolman, Harold 2–3, 4, 9–43
Wolverhampton, 13
women 54, 83, 84, 90, 92, 93, 96–8, 102–4
    passim, 107, 131–4, 138–9, 143, 151

Worcester 74

York 13, 17, 69, 71, 75
Yorkshire and Humberside 82, 104, 106, 123,
    126
Yorkshire, West 68, 106
Young, K. 170
young people 94, 95, 98–102 *passim*, 116, 121,
    122, 125, 148, 183–4
Youth Opportunities Programme (YOP) 110,
    116, 125, 135
Youth Training Scheme (YTS) *see* training